THE ENERGY IMPERATIVE

For decades, Hermann Scheer was one of the world's leading proponents of renewable energy. In this, his last book before his death in 2010, he lays out his vision for a planet 100 per cent powered by renewables and examines the fundamental ethical and economic imperatives for such a shift. And most importantly, he demonstrates why the time for this transition is now.

In Scheer's view, talk of "bridging technologies" such as carbon capture and storage or nuclear energy – even (and perhaps especially) by environmentalists – is actively damaging the more pressing agenda of the move to 100 per cent renewable energy. Instead, he offers up examples of the technologies which are working (economically) today and details the policy and market conditions that would allow them to flourish.

In 1993, Scheer's *A Solar Manifesto* laid the foundations for the road which has led to annual newly installed renewable capacity today rivalling that of conventional power sources. *The Energy Imperative* provides a practical, inspirational map for the next stage of the journey.

Hermann Scheer (1944–2010) was a Social Democrat member of the German Bundestag (Parliament), President of EUROSOLAR (The European Association for Renewable Energy) and General Chairman of the World Council for Renewable Energy.

THE ENERGY IMPERATIVE

100 Per Cent Renewable Now

Hermann Scheer

First published 2012
by Earthscan
2 Park Square, Milton Park, Abingdon, Oxon OX14 4RN

Simultaneously published in the USA and Canada
by Earthscan
711 Third Avenue, New York, NY 10017

Earthscan is an imprint of the Taylor & Francis Group, an informa business

This is an authorised translation of the work *Der Energethische Imperativ* by Hermann Scheer, published 2010 Verlag Antje Kunstmann GmbH, München.
Translation by Joanna Scudamore-Trezek.
Supported by the Hermann Scheer Foundation
(www.Hermann-Scheer-Foundation.org).

British Library Cataloguing in Publication Data
A catalogue record for this book is available from the British Library

Library of Congress Cataloging in Publication Data
Scheer, Hermann, 1944-2010.
[Energethische Imperativ. English]
The energy imperative : 100 percent renewable now / Hermann Scheer. – 1st ed.
p. cm.
Includes bibliographical references and index.
TJ808.S328713 2011
333.79'4–dc23
2011027112

ISBN: 978-1-84971-433-4 (hbk)
ISBN: 978-0-203-14414-5 (ebk)

Typeset in Sabon
by Taylor & Francis

Printed and bound in Great Britain by the MPG Books Group

FÜR LILLI SCHEER
GEB. 21.2.2004

CONTENTS

ACRONYMS AND ABBREVIATIONS

AKN	North Sea Action Conference (Aktionskonferenz Nordsee)
BauGB	Federal Town and Country Planning Code (Baugesetzbuch)
BDEW	Association of Energy and Water Industries (Bundesverband der Energie- und Wasserwirtschaft)
BEE	German Renewable Energy Federation
BGR	Federal Institute for Geosciences and Natural Resources (Bundesanstalt für Geowissenschaften und Rohstoffe)
BNatSchG	Federal Nature Conservation Act (Bundesnaturschutzgesetz)
BNetzA	Federal Network Agency (Bundesnetzagentur)
CCR	Carbon Capture and Recycling
CCS	Carbon Capture and Storage
CDM	Clean Development Mechanism
CDU	Christian Democratic Union (Germany)
CSP	Concentrated Solar Power
CSU	Christian Social Union (Germany)
DEC	Digital Equipment Corporation
DIW	German Institute for Economic Research (Deutsches Institut für Wirtschaftsforschung)
DPG	German Physical Society (Deutsche Physikalische Gesellschaft)
EBRD	European Bank for Recovery and Development
ECB	European Central Bank
ECF	European Climate Foundation
EEG	Renewable Energy Sources Act (Erneuerbare-Energien-Gesetz)
EIB	European Investment Bank
EU	European Union
Euratom	European Atomic Energy Agency
EWEA	European Wind Energy Association
FAO	Food and Agriculture Organization (UN)
FDP	Free Democratic Party (Germany)
FEMA	Federal Emergency Management Agency
FVEE	German Renewable Energy Research Association
GNP	gross national product

HVDC	high-voltage direct current
IAEA	International Atomic Energy Agency
IEA	International Energy Agency
IK	Informationskreis KernEnergie
IMF	International Monetary Fund
INFCE	International Nuclear Fuel Cycle Evaluation
IPCC	Intergovernmental Panel on Climate Change
IRENA	International Renewable Energy Agency
IRES	International Renewable Energy Storage Conference
ISET	Institute of Solar Energy Distribution Technology
KfW	Reconstruction Credit Institute (Kreditanstalt für Wiederaufbau)
KWKG	Combined Heat and Power Act (Kraft-Wärme-Kopplungsgesetz)
NATO	North Atlantic Treaty Organization
NEA	Nuclear Energy Agency
NGOs	non-governmental organizations
NRDC	Natural Resources Defense Council
OECD	Organisation for Economic Co-operation and Development
OPEC	Organization of the Petroleum Exporting Countries
OPURE	Open University for Renewable Energy
PIK	Potsdam Institute for Climate Impact Research (Potsdam-Institut für Klimafolgenforschung)
REDD	Reducing Emissions from Deforestation and Degradation
ROG	Regional Planning Act (Raumordnungsgesetz)
RWI	Rheinisch-Westfälische Institut für Wirtschaftsforschung
SPD	Social Democratic Party (Sozialdemokratische Partei Deutschlands)SRU German Advisory Council on the Environment (Sachverständigenrat für Umweltfragen)
UN	United Nations
UNCED	United Nations Conference on Environment and Development (Rio Summit)
UNEP	United Nations Environment Programme
UNESCO	United Nations Educational, Scientific and Cultural Organization
UNSEGED	United Nations Solar Energy Group on Environment and Development
WBGU	German Advisory Council on Global Change (Wissenschaftlicher Beirat der Bundesregierung Globale Umweltveränderungen)
WTO	World Trade Organization

FOREWORD

Bianca Jagger

Two revolutions have shaped the course of recent human history: the industrial revolution and the information technology revolution. Hermann Scheer was a driving force behind the third significant revolution, the renewable energy revolution, which will determine the shape of our future. While Hermann was writing this book, he told me that this would be his definitive work, which I believe it is. He was full of enthusiasm when he finished it, exhilarated. He felt that he had succeeded in writing a book which would provide significant solutions to the challenges we face in the world today.

I was fortunate enough to know Hermann as both a friend and colleague, and to have the privilege of working closely with him over the years. In 2007 I founded the Bianca Jagger Human Rights Foundation. Hermann served on the Board of Advisors. His insight was invaluable, and his expertise and generosity unparalleled. He was a guiding light, and is greatly missed. I share Hermann's belief that climate change is not just an environmental issue, it touches every part of our lives: peace, security, human rights, poverty, hunger, health, mass migration and economics. It is a global issue, and it calls for global action. As Hermann says in *The Solar Manifesto*,

> To be able to discuss energy as a separate matter is an intellectual illusion. The CO_2 emissions are not the only problem of fossil energy. The radioactive contamination is not the only problem of atomic power. Many other dangers are caused by using atomic and fossil energies: from the polluted cities to the erosion of rural areas; from water pollution to desertification; from mass migration to overcrowded settlements and the declining security of individuals and states. Because the present energy system lies at the root of these problems, renewables are the solution to these problems.

When speaking about climate change and renewable energy, I often reference Hermann's work, in particular his excellent books, *The Solar Economy* and *The Solar Manifesto*, in which he says that the transition to renewable energy is the over-riding moral imperative of our time. He states that renewable energy costs will generally go down, as they largely consist of technology expenses. Mass production and technological innovation will bring dramatic decreases in cost. He was emphatic that we should see the promotion of renewables not as a burden, but as a unique economic opportunity. Thanks to Hermann, exemplary progress has been made worldwide towards a green electricity supply from renewable sources rather than coal, gas and nuclear power.

Hermann was a pioneer. His leadership in the field of renewable energy, and his passionate commitment to the environment and to human rights was a constant inspiration. *TIME* magazine called him a "hero for the green century" and cited his "radical vision" and "gale-force enthusiasm",[i] which "electrified Washington audiences". These are well deserved accolades for a man who devoted his life to making a difference in the world. His many significant accomplishments speak for themselves. He introduced feed-in tariffs, otherwise known as Scheer's law, during his time as member of the German Parliament. Although feed-in tariffs had previously existed, on a smaller scale, in Denmark, what Hermann achieved in Germany is unique. His implementation was rigorous in ensuring that the price paid for renewable energy gave the producer an adequate return, which meant that it was a worthwhile investment. Thanks to Hermann's vision and hard work, feed-in tariffs have become a landmark piece of legislation, incentivising the use of renewable energy all over the world.[ii]

Hermann's stance was bold, and uncompromising. Other environmental experts continued to advocate a partial conversion to renewable energy, claiming that electricity systems must continue to rely in part on coal, nuclear and fossil fuels. Hermann was adamant that we should aim for a 100 per cent conversion to renewables. His conviction that this was not only possible but necessary has been borne out by the growth and success of the sector in recent years. Hermann knew that renewable energies can stimulate technological innovation and economic development, and that renewable energy will become fully competitive with conventional energy systems. Onshore wind power has already achieved parity with coal-fired power, while being much less environmentally damaging. Hermann understood the crucial economic and social benefits renewable energies could have for the poorest countries in the world. Homegrown renewable sources can help developing countries to fuel their economic development and to insulate themselves against rising world energy prices.

The governing principles behind Hermann's approach were the democratization and decentralization of energy. Energy autonomy, for countries and even for individuals, was his fundamental goal. He believed that, by producing as much energy as possible locally, we could reduce global dependence on long distance transmission lines, and diffuse the concentration of economic power, which resides largely with a very few companies and institutions. Hermann was an anti-monopolist. He knew, too, that renewable energy doesn't necessarily automatically lead to democratization. His principles were rigorous, his warnings against projects such as the proposed DESERTEC supergrid, intended to facilitate electricity transfer from North Africa to Europe, were clear.

"The countries in which these installations are located do not gain adequately from them", he wrote to me in an email of 6 July 2009. "They need decentralised energy production and supply even more badly than we do because their grid infrastructure is less developed than ours. The example of oil companies in the oil exporting countries are a warning. Those mega-investments hardly ever reach local populations – instead they often reinforce local disparities."

Hermann was an innovator, a passionate advocate of renewable energy before it was recognised, or taken seriously. He was a tireless campaigner for solar power; he never allowed criticism to deviate him from his objective. Some accused him of pursuing a utopian paradigm, yet as time goes on and the impending global energy crisis looms ever larger, Hermann's vision seems prophetic.

Hermann's commitment to renewable energy was unwavering; he was head of the European Association for Renewable Energy (EUROSOLAR), and Chairman of the World Council for Renewable Energy, WCRE, two non-profit international organisations, and in 1999 he was awarded the Right Livelihood Award, otherwise known as the alternative Nobel Prize. He was the driving force behind the establishment of the International Renewable Energy Association, IRENA. He realised that promoting renewables must become a global and universal priority, and that the creation of IRENA was a necessary condition for that goal. Together with EUROSOLAR and the WCRE, Hermann struggled for two decades to make IRENA a reality. It exists today because of his powers of persuasion, his consuming passion and his hard work. He knew that there could be no global renewable energy revolution without IRENA. As Hermann said in his speech at the founding conference on 26 January 2009, "Rapid action is indispensable. The time for paying lip service to renewable energies is over. An end to the game of 'talking globally, postponing nationally' is well overdue".[iii]

Hermann Scheer has been called the "solar king", the "sun god", "solar crusader", and the "solar pope";[iv] he was all those things and more. Hermann was a visionary, who set us on the path towards a world without fossil fuels, coal and nuclear power. He paved the way for "a new generation of decision makers ... in the established energy sector that recognizes that nuclear energy and fossil-fuel energy lead to a dead end".[v] Hermann was a force for change, a voice of reason in a sector dominated by big business, lobbyists and spin doctors. His work has helped to shape the way we perceive and implement renewable energy. As the *TIME* article goes on to say, "Members of the German parliament don't usually create much of a stir outside Europe. But Hermann Scheer ... has, more than any single political leader, transformed Europe's energy landscape".[vi]

Herman Scheer's passing is a great loss. I have many fond memories of his *joie de vivre*, his great vigour, and his infectious sense of humour. He was uncompromising when it came to his principles, a staunch and loyal friend, a courageous campaigner, and a man of unshakeable integrity. He has left an enduring legacy behind and is greatly missed.

Bianca Jagger
20.09.11

Notes

i www.time.com/time/magazine/article/0,9171,1003146,00.html#ixzz1XjP1qcYO
ii www.guardian.co.uk/world/2010/oct/18/hermann-scheer-obituary
iii www.wcre.de/en/index.php?option=com_content&task=view&id=99&Itemid=3
iv www.guardian.co.uk/world/2010/oct/18/hermann-scheer-obituary
v www.eurozine.com/articles/2011-04-22-scheer-en.html
vi www.time.com/time/magazine/article/0,9171,1003146,00.html#ixzz1XjMBSiNq

INTRODUCTION

Energy change: The ultimate challenge

These days, the whole world talks of renewable energy, as happily as it does about good weather. Scarcely anyone continues to deny that the future for mankind's power supply lies in renewable energy. For a long time this idea was regarded as fantasy.

This change in perception goes back only a few years. In May 2002, the United Nations (UN) invited me to join a small and select group of people at a meeting in their New York headquarters, to help clear up a problem which had occurred to Kofi Annan, then Secretary-General of the UN. The UN was in the last stage of its preparations for the World Summit on Sustainable Development which was subsequently held that August in Johannesburg. But in the draft for the final declaration, drawn up during several preparatory conferences, there was not a single mention of the key role played by renewable energy for the sustainable development of world civilization. It was our task to suggest formulations to fill this gap. The episode shows just how deep and widespread ignorance of renewable energy still remained at the beginning of the 21st century.

The current global attention being paid to renewable energy arose outside mainstream political, economic and media discussions on energy. These remain ensnared in a global view of power supply dominated by nuclear and fossil fuels. In the 1990s, the few pioneers of a 'solar age', which relies neither on nuclear nor fossil energy, still came up against deep-seated psychological, and enormous practical, barriers. Today these appear to have been overcome, although more in words than in thoughts and deeds. Wholehearted declarations from governments and power companies which engender the impression of absolute commitment to sustainable energies blur the view for the practical priorities. Although now power companies too are investing in renewable energy, their focus remains first and foremost on conventional energy sources – where possible right down to the last drop of oil, the final ton of coal or uranium and the last remaining cubic metre of gas. These are of greater value to power companies, for wind and solar radiation cannot themselves be sold as resources. Resistance to renewable energy has given way to a strategy of monopolization and delay, a strategy which demands that the increasingly urgent transition to renewable energy be introduced only in cautious and thereby often questionable stages.

Even so, in the meantime it is universally recognized that the future of power supply must lie in renewable energy. The many dangers and limits associated with the extraction and production of nuclear and fossil fuels have become manifest. For this reason alone, renewable energy can no longer be ignored, particularly in view of its impressive growth rate. Between 2006 and 2008 alone, global annual investment in renewable energy doubled from US$63 billion to US$120 billion. The worldwide installed capacity of wind power plants grew from 74,000 to 135,000 megawatts between 2006 and 2009, and that of photovoltaic plants connected to networks from 5,100 to 19,000 megawatts. Cracks in a world view based on nuclear and fossil energy began to appear with the admission that renewable energy offers the potential for global application. Its psychological strength derives from the realistic hope it offers of safe and secure energy supplies over the long-term. Thus, renewable energy represents a superior social value over nuclear and fossil energy. This is the crux of the matter when it comes to energy considerations.

Those who identify renewable energy not only as a supplement to current energy supplies, but also a tangible, comprehensive alternative, can hardly continue to deny this. Given a real choice, most people will opt for renewables over nuclear and fossil energy. Germany is the practical example of this. After the German government's Renewable Energy Sources Act (EEG) came into effect in 2000, and despite continued resistance, by 2009, the share played by renewables in generating electricity had grown from 4.5 per cent to 17 per cent, and in overall energy supply from 3 per cent to 10 per cent. People's trust in the potential offered by renewable energy grew parallel to this development and with it the hope and expectation of being able to rely completely on renewable energy in the near future. Surveys indicate that 90 per cent of Germans favour the further, extensive exploitation of renewable energy – with 75 per cent wishing to see this in their local regions – and they are even prepared to accept higher energy costs to make this possible. Fewer than 10 per cent favour the building of new nuclear or coal-fired power plants.[1]

The widespread popularity of renewable energy has developed despite decades of extensive public denunciation by the traditional power industry and the majority of energy experts. It has been driven – and notoriously is still being driven – by huge publicity expenditure. Even so, this duo of opponents is losing ground in the battle for public opinion, a battle that remains equally intense, although now fought by more subtle means.

Today the discussion primarily revolves around the question of the time needed for a complete transition to renewable energy. Is this only possible after 2100? Or can it be achieved by 2050? I am convinced that this transition can be realized faster if we mobilize all the necessary strengths; globally we need a period of around a quarter of a century, and in several countries and regions we can do this faster. It is not only the enormous natural potential offered by renewable energy which makes this transition possible – we also have the necessary technological potential. We need to make this change, both for ecological reasons and because we also clearly need to secure our economic existence. Such a transition is not an unbearable strain, but rather a new, global economic opportunity for the industrialized countries, and *the* major chance for developing countries. However, the greatest potential for change is embodied by mankind itself: making ourselves and, above all, "politics" and "the economy", into active proponents of renewable energy is the key to success. This demands a unique political and cultural show of strength. Moreover, the challenge we are facing is also historically unique. The longer we procrastinate, the harder it will be to overcome this challenge. We have already lost too much time.

Why, when and how?

If the transition from nuclear and fossil to renewable energy is only carried out in a piecemeal and gradual manner, then it is highly likely that world civilization will be thrown into a staggering crisis affecting everyone and everything: dramatic climate change threatens to make entire habitats unfit to live in and to trigger mass misery and the migration of hundreds of millions of people. This burdens civilization with challenges and damage limitation costs greater than those required for the transition to renewable energy. Already the reduced availability and increased costs of power supplies generated from nuclear and fossil resources are having dramatic economic consequences, causing social fractures in the industrialized countries and allowing developing countries to become ever more impoverished. We face the threat of growing international conflict over access to remaining resources, the severity of which could lead to war. The problems associated with nuclear power – from constant security threats to ongoing operations, through to nuclear terrorism and the thousand year "writing on the wall" for nuclear waste – remain both unsolved and unsolvable. The huge volumes of water consumed by nuclear and fossil power plants aggravate the water crisis in a growing number of regions. The dangers to health posed by nuclear and fossil-based energy supplies are increasing, and the contamination of marine life by oil extends into the food chain. All these crises, appearing simultaneously and exacerbating one another, are cutting societies to the quick. They are a greater threat to the industrialized model of civilization, one based on fossil and nuclear energy, than the global financial crisis. Whether in its guise as a capitalist, market economy or a socialist, planned economy, this model of civilization has already severely damaged the foundations of our existence.

Therefore every year wasted in making the comprehensive and complete transition to renewable energy is a year lost. This change is the *ultima ratio*: the last possible means of averting existential and possibly irreversible dangers. It is of ultimate significance because there is no other way for humans to generate energy in a natural and sustainable manner. The consequences of the established power supply system compel us to urgent and decisive action.

Clichéd declarations of commitment to renewable energy reveal little about the significance this commitment is actually awarded: first, second, or third rate? For whom are such declarations simply a concession to a worried public? Have all those who denied renewable energy for so long really been converted? Will change be presented as compulsory or seen as deferrable? It was Mahatma Gandhi who said: "First they ignore you, then they laugh at you, then they fight you, then you win." In which of those first three phases we currently find ourselves varies according to country and the status of its public debate and development. More than half of the world's installed wind power generating capacity is based in only six countries (the US, Germany, China, Denmark, Spain and India). Around half of the world's photovoltaic plants connected to electricity grids are installed in Germany alone. More than 80 per cent of the installed capacity for solar thermal energy supplies is concentrated in China and the countries of the European Union (EU). Obviously, far too many people in too many countries continue virtually to ignore renewable energy.

Some excuse their reluctance by claiming that the shift to renewable energy takes "much time" and that overly ambitious and rapid steps in this direction represent an unreasonable strain on the economy. Some truly believe this, others use this as an excuse to gain time and carry on as usual for as long as possible. Some lack the courage to break with the structures of traditional energy supply; others are helpless and have no concept of how to implement energy change. Good intentions do not empower one to act; rather they are simply a prerequisite to action.

At least the times are past in which public declarations that renewable energy offers a global alternative to nuclear and fossil energy would be followed by a fierce barrage of criticism. The first two phases of Gandhi's process by which an idea becomes reality – being ignored and ridiculed – have already been used as targeted methods of fighting renewable energy. Today the question of *whether* renewable energy alone is able to supply all our energy needs is categorically and overwhelmingly answered with "yes". As a result, many have been led to believe that the conflict over renewable energy has abated and that a fundamental consensus has been reached. From now on we are "merely" dealing with the *when* and the *how*. However, this raises key questions:

- Which traditional energy sources, i.e. nuclear power and/or fossil energy, should be relied upon until all our energy needs can be met by renewable energy alone?
- Which of the options offered by renewable energy should be chosen? How can they complement one another? How great is our real energy storage need?
- What structures should be used to make renewable energy available: decentralized and/or centralized?

- Which political policies are crucial for making the general transition to renewable energy? Should the focus be local, national or international?
- Which stakeholders can drive forward energy change and what is the role of the traditional power industry?

The answers to these questions are politically and economically explosive. They play a key role in determining the answer to the most important question – the *time question*. Can a complete transition to renewable energy, historically due, be realized fast enough for us to still escape the disasters caused by traditional energy supplies? Who and what is applying the brakes, and how can we accelerate this process? It is this last question against which all previous questions will be measured.

The illusion of consensus

The implied consensus on renewable energy distracts us from the fact that the true conflict has just begun, although on a different front. This consensus seduces us into underestimating the inevitable conflicts associated with energy change, which means we fail to face up to them. They differ from earlier conflicts over renewable energy and are also more profound. Where the transition to renewable energy has, to all intents and purposes, begun, we have now arrived at the moment of truth: the effective substitution of nuclear and fossil energy by renewable energies has a direct effect on established energy structures. These are closely intertwined with the dominant systems of production and consumption, economic structures and political institutions. This shift has an immediate impact on the existential interests of the traditional power industry, the largest, and above all politically most influential, sector of the global economy. This is reflected in the conflicting developments in global energy activities.

The first steps towards the breakthrough of renewable energy as outlined above are being matched by political initiatives, such as those made by US President Obama, the Chinese and Indian governments, and even of oil and gas exporting countries in the Gulf region. The EU has now enacted a law requiring that, from 2012 onwards, all future public buildings meet zero-emissions standards and, from 2020, all private buildings do so too – a target only achievable using renewable energy and energy-efficient construction methods. China is training 10,000 solar energy technicians in Africa. In Bangladesh, more than 100,000 small solar power plants are being installed each year, thanks to microcredit and the training of technical service personnel. In the meantime, in Germany, whose EEG made it an international pioneer in producing electricity from renewables during the first decade of this century, all the political parties are talking of converting the entire power supply to renewable energy by the middle of this century. Numerous German towns and districts have decided to completely convert to renewable energy within

10 to 15 years; some have already managed this for the production of power and heat, as have some local authorities in Austria and Denmark.

Large international corporations including Bosch, General Electric and Siemens have made renewable energy a strategic focus. Power companies such as Germany's E.ON and RWE are making significant investments in renewable energy. Automotive companies are gearing up to manufacture electric cars and favour meeting power needs with renewable energy. Major banks, from New York's Wall Street to the banking centres of Frankfurt and London, have drawn up impressive renewable energy portfolios, and investment funds for renewable energy are mushrooming. The majority of small and medium-sized companies who specialized in renewable energy and became pioneers are no longer alone. Some are growing; others are being taken over by large corporations in an effort to catch up on lost time.

Yet on the other hand, there is an unmistakeable growth in countertrends which continue to reveal completely different priorities: increasingly significant sums are being invested into conventional energies on a global scale – in 2009 they went up four-fold. These are investments in conventional large-scale power plants and pipelines, some worth tens of billions and with long depreciation times, thereby serving to solidify the status quo for several decades to come. In order to push through his own initiatives, President Obama was forced by the US Congress to concede the continued promotion of nuclear power, the construction of new coal-fired power stations and controversial authorisations for oil drilling and pipelines. China and India are focusing on constructing new coal-fired power stations. Billions in subsidies have already been sanctioned for so-called Carbon Capture and Storage (CCS) power stations, designed to capture the CO_2 released in coal-fired power stations for eventual underground storage. Already the European Commission has allocated more investment funding to this technology than it has made available as direct investment into renewable energy. Energy giant Shell has largely abandoned the solar initiative it began in the 1990s, announcing instead that it will invest in CCS. In the Canadian province of Alberta, gigantic excavators dig up the tar sands in an excavation area extending over 20,000km to produce fossil energy. This is accompanied by alarming invasions of the natural water table, with 20 litres of water needed to produce a single litre of oil. In West Virginia, the coal region of the US, entire mountains are demolished to extract ever larger quantities of coal using ever bigger excavators. The whole world followed the catastrophic consequences of deep sea drilling on the seafloor of the Gulf of Mexico, near the US coastline. Already some are hoping that the ice at the North Pole will melt, enabling the fossil energy resources lying beneath to be extracted.

France's President Sarkozy has introduced more initiatives for renewable energy than his predecessor. However, his primary role has been that of an international travelling salesman, winning contracts for French companies to build nuclear power plants abroad. Although in April 2010, the British government passed a law on feed-in tariffs for renewable energy based on the German model, it is simultaneously building new nuclear power plants. Also in April 2010, the Finnish government authorized the building of two new nuclear reactors even though the Green party is a member of the government. Italy's government, led by Premier Berlusconi, has announced the building of nuclear reactors even though, in a 1987 referendum, the population rejected nuclear power plants. In the spring of 2010, Russia and Ukraine agreed a joint plan to bundle their atomic know-how, to double the production of electricity generated by nuclear power within a decade, and to offer this power on the international market. At the beginning of 2010, Abu Dhabi ordered four nuclear reactors from Korea, and Vietnam wanted to start producing nuclear power. Even Brazil, who together with Russia, Canada and Australia possesses the greatest natural sources of renewables, is planning to build new nuclear power plants. The International Energy Agency (IEA) demands that 32 new nuclear power plants be built each year, right up to 2050, which would mean a new one roughly every 11 days. This is an aggressive policy of carrying on as usual, even when it has been proven that these energies are more expensive than renewables – everything we find must be excavated and sold. The global "pyromania", as I call this compulsive obsession in my book *The Solar Economy*, continues assiduously. "Drill, drill, drill," is Big Oil's rallying cry in its war on the environment, a cry not even muted during the drilling catastrophe in the Gulf of Mexico. This fateful game with our planet continues, always justified with the claim that the "current" potential offered by renewable energy is insufficient.

Thus, so the thinking that, although renewable energy has become socially acceptable, it should in no way affect the continued existence of fossil and nuclear energy, and be available only to satisfy additional energy requirements – there should be as little substitution of nuclear and fossil energy as possible! This interest in maintaining the status quo is also being asserted in Germany, reflected in attempts to reverse the 2001 agreement on the phase-out of nuclear power and in the numerous plans for constructing new coal-fired power stations which boast capacities which assume that renewables cannot play a greater role. This is accompanied by an increasing number of orchestrated attacks on Germany's EEG which stimulates the rapid expansion of renewable energies. Scientific institutions, too, have joined the groups actively playing down the potential offered by renewables. The new consensus on renewable energy is illusory. The established powers in the power industry are focusing on, at best, the coexistence of nuclear and fossil energy with renewables, awarding the greatest possible role to the former and demanding that renewable energy adapt to the structures of traditional energy supply, being channelled and restricted accordingly.

So the fundamentals of the energy conflict have hardly changed. It was only ever superficially about the pros and cons of renewable energy – the core of the conflict has always been the structures of power supply and the authority to dispose over them. The focus on fossil, and later nuclear, sources of energy, has led to the creation of the current power supply system; a reorientation towards renewables endangers its very structure. Therefore, after the phases of rejecting and belittling the pioneers of renewable energy, the powers-that-be are now focusing their efforts on curbing the speed of energy change. Consequently, attempts by the traditional power industry to influence political decisions, the media and public opinion, are also increasing. In the US, immediately after President Barack Obama stepped into office, more than 2,000 additional, highly paid lobbyists for the American power industry were sent to Washington with the task of blocking the announced energy turnaround through the targeted handling of members of Congress and the media. Large sums have been spent on "greenwashing", the term used by German journalist and author Toralf Straud in describing the efforts made by business and industry to appear green.[2]

The phenomenon of former government members inconspicuously slipping into management roles at power companies immediately after leaving office is increasing, as is the number of journalists being employed by power companies as media consultants to tend to their public image. Shrouded by the secondary business of renewable energy, the primary business of conventional energies carries on as usual.

The *nervus rerum*

The "new" energy conflict erupts primarily where the introduction of renewables is sufficiently advanced as to be able to replace a significant share of conventional energies. The true problem behind the illusion of consensus – the *nervus rerum* of the conflict – is that the systemic requirements (i.e. the overall technical, infrastructure, organizational, financial and, last but not least, political input) of nuclear and fossil energy are incompatible with those of renewables. However, our goal must be to dismantle the traditional energy system completely. To assign only a limited role to renewables would be a strategically unjustifiable self-restriction. It would mean being forced to continue with conventional systems and to support them politically over the long-term. Thus we end up with two opposing systems of power supply which, once a certain point has been reached, would hinder one another's progress.

Doubtless we must pass through a transitional phase on the route to 100 per cent renewable energy, with the growing share played by renewables being matched by a corresponding shrinking of the role of conventional energies until these are finally, and completely, replaced. However, during this phase the decisive factor will be the system requirements that dominate – will they be those of the established energy system or those suited to renewable energy? Thus conflict is inevitable, one unique in the history of modern power supply. On the one side we have traditional sources of energy, with the entire power supply system structured according to their functional requirements and supported by tailor-made laws. On the other, we have the prospect of a system based completely on renewables, and with largely contrasting functional requirements for which there have, so far, been only attempts to develop a political framework.

Between the current status and the status to which we aspire lies a phase marked by friction and discord. Let's call it a *hybrid phase*, analogous to the hybrid car which is equipped with two engines, each running on a different type of fuel. The trump card held by the traditional energy system is that it is well established. Consequently, it demands that energy change be slow and take place according to its own rules. In contrast, the strongest suit held by renewable energy is not only the lack of any alternative, but that it is increasingly used independently from the traditional energy system and is more highly valued by society. However, currently we still find ourselves in a state of trial and error, with a multitude of competing concepts, some more, some less well thought-through, and thus easily played off against one another. This is the true problem in bringing about energy change.

The key question is how to overcome this obstacle and expand the use of renewables quickly. While it is vital to recognize both the weaknesses and strengths of the established energy system, every strategy for implementing renewable energy must be based on, and accentuate, the intrinsic strength of renewables. These strengths and weaknesses are of a technological, economic, psychological and, last but not least, political nature. As this systemic tension is the *nervus rerum* of energy change, this is the conflict on which we focus in this book.

Old and new fronts

During this hybrid phase of restructuring, both the fronts and stakeholders change. For a long time, the dividing line between renewable energy advocates and supports of conventional energy was clear cut and straightforward. On the one side, an initially small number of renewable energy pioneers, including renewable energy organizations, environmental alliances, environment institutes, political figures, pioneering companies and sympathizers in the media. On the other, an almost unanimous group of renewable energy rejectionists, drawn from the energy industry and the governments with whom they have always worked closely, established research institutes, trade associations, and most industrial enterprises and business media. In the meantime, the fronts have softened and some of the stakeholders have changed sides.

Industrial enterprises, financial institutions and investment groups have recognized that building plants and equipment for renewables and financing renewable energy projects are attractive and profitable business propositions. For a long time, the business and industrial enterprises stood arm-in-arm with the established power industry, joining them to denounce renewable energy as unfit and accuse its proponents of hostility to business. Now they sing the praises of the market opportunities generated by renewables. Municipal power companies, who have long become appendages of traditional power suppliers, see in renewable energy the chance for future independence. The more popular renewable energy becomes, the more political parties and institutions are willing to engage with it.

A new generation of decision-makers is also emerging in the established power industry. Recognizing that nuclear and fossil energy offer no future, they are attempting to make the move towards renewables in a way that fits with the structures of established power supply. Past methods of rejection have been exhausted – now it's all about joining in and jumping on the bandwagon, in an attempt at least to influence its destination and speed. Moreover, power companies are trying publicly to justify their adherence to nuclear and fossil energy by investing in renewable energy.

While the front led by the former defendants of conventional energies has collapsed, the spectrum of renewable energy advocates has become more differentiated. Political movements that have encouraged this development need to be modified but, although there are many suggestions as to how to do this, they often lack consistency and clarity of vision. Competing interests, which have emerged as renewable energy has developed, erupt as soon as it comes to grabbing shares of a growing cake. Advocates of renewable energy who nonetheless still believe in a pivotal role for the established energy industry interpret the change of tone as traditional energy's willingness to cooperate. Manufacturers of renewable energy plants and equipment are receiving orders from power companies and becoming business partners. Research institutes for renewable energy are now also being contracted to carry out research on behalf of traditional power companies. Governments host discussions designed to achieve consensus on joint efforts between conventional and renewable energy and the mutual staking of claims. Many advocates of renewable energy, long despised as outsiders, see this as a major step forward. Yet as consensus is always preferable to conflict, this results in a willingness to compromise, one in which an invisible limit is often suddenly and inadvertently crossed, where comprise stops and being compromised begins.

All this is typical for transition periods in which all parties adapt to a new situation and many hope for a consensus in order to gain a certain security. Not everyone is able or willing to think about the overall developments. As helpful and constructive as consensus can be, it can also be most crippling. The question must always be: consensus about what and with whom, and who has the upper hand? Consensus among everyone affected by change, in whatever form, inevitably slows progress. Or consensus among those who join forces in striving for a joint goal? Shared agreement on rapid energy change is only conceivable where the goal being pursued offers the prospect of a win-win situation for all involved. This is a promise happily uttered by those wishing to avoid the necessary conflict. Seen objectively, however, in the transition to renewable energy there can be no win-win situation.

The transition to 100 per cent renewable power supply represents the most comprehensive economic restructuring since the beginning of the Industrial Age, and it is inconceivable to imagine that this process will not have both winners and losers. The losers will inevitably be the traditional power industry, with the extent of their loss depending on their insight, on their willingness and ability to engage in root and branch reform, to face up to drastically sinking market shares, and to discover new fields of entrepreneurial activity outside the power industry. Attempts to escape the role of loser in this process of change and to hang on to a central role in the energy business leads to contradictory, ineffectual and expensive slowing-down strategies. The winner of this change will be world civilization as a whole, its societies and economies, technology companies and many local and regional businesses. Whatever the outcome, energy change will have significantly more winners than losers. Many potential winners are not yet aware of the opportunities and so remain on the opposing side. The current incumbents are the potential losers and currently have more influence on actual events, while the potential winners, by no means established, have far less.

Real realism

It is true that every economic or political initiative to promote renewable energy drives development forward in some way, irrespective of the underlying motives. However, they are neither of equal value and nor are they equally suited for effecting rapid energy change. Therefore it is vital to separate the wheat from the chaff and to recognize:

- which initiatives facilitate the unlimited development of renewable energy and which permit this only to a limited extent – and do these initiatives complement or hinder one another;
- which policies encourage growth in the number of renewable energy supporters and provide them with the necessary room for manoeuvre, and which have the opposite result of reducing the spectrum to just a few players upon which future progress then depends;
- which initiatives do justice to the multitude of reasons for the transition to renewable energy, and which focus only on a single goal (e.g. climate protection), thereby limiting their overall efficacy.

The catalogue of contentious questions is extensive: what should be made dependent on international agreement? Are global climate negotiations the silver bullet upon which everything else depends, or are they a well-trodden path along which there is almost no chance of making headway? Does emissions trading encourage energy change or act as a brake? Do we need more comprehensive, multilateral approaches or more individual pacesetters? What value do the various options represent for renewable energy? Should renewable energy be appropriated predominantly in those regions where more sun shines or more wind blows, or everywhere? What do we mean by “cheap” and “economic” power supply? Increasingly the most contentious question is that of “decentralized” or “centralized” renewables-based power supply structures: are large-scale power plants necessary at all? If so, under what conditions? Is it also essential to expand power supply networks on a major scale by means of "super grids" even where the provision of renewable energy is predominantly decentralized, or should the focus be on regional and local "smart grids"? These questions lead not only to differing plans of action but also to conflicting objectives for introducing renewable energy which need to be discussed and battled out. However, political parties and governments, as well as many advocates of renewable energy, shy away from these questions and, in order to keep the peace, declare all these diverging policies to be equally important and worthy of promotion.

Disputes about the ways and means of achieving the transition to renewable energy are being fought out in political institutions as well as in environmental organizations and those promoting renewable energy. These disputes are confusing many people and generating public and political uncertainty about the necessary route to achieve energy change. Therefore we need to take stock. A critical analysis of the various approaches, one that examines their likelihood of practical success and their consequences, is long overdue. The main question is why the obvious and urgent matter of our energy survival, also an ethical concern, is still largely being dealt with in a half-hearted manner. The reasons given for this are specious and a forceful push towards energy change is now essential. There are longer and shorter "roads to Rome", paved with a multitude of various obstacles and challenges to implementation, and with a diverse range of political, economic, social and cultural ramifications. This makes it all the more important to recognize the fastest paths to energy change. Our decision whether to set off along these paths should reflect not just business management or energy policy considerations but also economic, political and, last but not least, ethical grounds.

The systematic and critical analysis of the current situation I attempt in this book is intended as a navigational aid for breakthrough strategies. It is based on my experiences in development and in implementing the most successful political initiatives for renewable energy at national and international level to date, as well as my observations on the success and failure of concepts for renewable energy in many countries. In order to make something happen, one must recognize the likely obstacles and understand how to overcome them. This demands that we understand the interests and intentions behind these obstacles and the strengths that are required to overcome them. Everyone involved needs to understand the realism upon which their action is based. Too many interpret this as the task of only pursuing that which seems possible under the current conditions and existing power structures. However, if analysis of the current situation indicates that the limited opportunities to act do not provide an adequate answer to the real challenges we face, then we need a different understanding of realism, one aimed at changing the parallelogram of forces in order to expand the room for manoeuvre. In view of the climactic dangers arising from established power supply systems, politics can no longer be regarded as simply the "art of the possible". Instead it must become the "art of the necessary". This is the *real realism* needed for energy change. Analyses and concepts must be thought through in an uncompromising manner. Although generally unavoidable, compromises should only be permitted during the actual process of implementation. Therefore I show the bottlenecks that need to be overcome and explain why one-dimensional observations lead us nowhere. Opening up our minds is the prerequisite for practical breakthroughs.

The political key to energy change lies in breaking up the established framework in which the power industry operates, a framework which is perforce limited, constricting the global economic, social and cultural opportunities for the transition to renewable energy. Energy change is a macroeconomic and social project which is focused on the future. It cannot be realized using only the methods and calculations of the traditional power industry. The range of technological capabilities is constantly growing, making change possible at a speed that pragmatists who focus on the present believe impossible. Rapid energy change needs multitudes of autonomous activists, prepared neither to hold back with their initiatives, nor to wait and see what others do. This thesis, which I set out and explain in my book *Energy Autonomy* (2005), has now been confirmed in practice, in the face of the prophecies of doom issued by the usual experts on practical constraints. Politics needs to forge new pathways to overcome this faintheartedness, not least because the traditional power industry has only been able to secure and maintain its dominant role thanks to various forms of political patronage. This patronage, rarely mentioned and examined far less critically than renewable energy initiatives, must be denounced politically.

Recognizing the fundamental importance of energy change for society's ability to survive in the future, my starting point is not renewable energy but rather society itself. I have not moved from renewable energy into politics in order to implement them. Rather, it is my view of the fundamental problem and my understanding of political responsibility that has led me to renewable energy. The transition to renewable energy is of historic significance for civilization and we need to know how to speed this process up. It is not renewable energy that we lack, it is time.

Part I

Taking stock

1

NO ALTERNATIVE TO RENEWABLE ENERGY

The long suppressed physical imperative

How is it possible that we have reached such a dramatically decisive point, one at which our very existence is endangered if we fail to make the urgent transition to renewable energy? Why has renewable energy been opposed for so long and valued so little? Despite Albert Einstein claiming to be "more interested in the future than the past, for it is in the future that I intend to live", these questions still need to be asked. Yet to put us on the right track for dealing with the future, we should also note philosopher Søren Kierkegaard's dictum that "life can only be understood backwards, but it must be lived forwards". Looking back helps us to clearly grasp the mental and structural obstacles to future developments. Whether we are aware of them or not, every past leaves its traces, be they philosophical, physical or psychological. What is remarkable is not the recognition of the fundamental value of renewable energy, which has grown in leaps and bounds over the past few years, but how long this process of recognition has taken. It is also noteworthy that so few tried and tested renewable energy technologies and initiatives are currently in place.

According to the laws of physics, it was always inevitable that the exploitation of fossil energy could only ever be a transitional stage. Wilhelm Ostwald, winner of the 1909 Nobel Prize for chemistry, points out starkly and irrefutably in his 1912 book, *The Energetic Imperative,* that "the unexpected legacy of fossil fuels" leads us into "losing sight temporarily of the principles of a durable economy and into living from one day to the next". As it was unavoidable that these fuels would someday be exhausted, we are forced to recognise that a "durable economy needs to be based exclusively on the regular influx of energy from the sun's radiation". Thus, his imperative – "don't waste energy, utilize it". By "waste", Ostwald was referring to the combustion of fossil fuels. This is a destructive process because the resources used in generating energy are irretrievably lost. This is followed by his admonition to use energy which is always available – what we today call renewable energy – and what in Denmark is fittingly referred to as "enduring energy". Ostwald assigns to his *energetic imperative* a greater social value than philosopher Immanuel Kant's *categorical imperative* which states that one should "act as if the maxim of thy action were to become, by thy will, a universal law of nature".[3]

This is more simply expressed by the old proverb of "do unto others as you would have them do unto you". Everyone wants to breathe clean air, so no one should be allowed to pollute anyone else's air. Everyone needs energy resources, so no one should be allowed to use so much energy that none is left for others. What is certain, is that there have never been sufficient sources of fossil and nuclear energy to meet all mankind's energy needs, and in the near future this will be increasingly the case, as reserves are exhausted and demands for energy grow. Ostwald sees in Kant's imperative a moral law of behaviour, and in his own, a physical law. Whether a moral law is observed or not is a question of ethics and determines the quality of social cohesion. A physical law, in contrast, offers us no choice. Failing to observe this physical law has such devastating consequences for society that it would also, in the end, make it impossible to realize Kant's ethical maxim.

Although an internationally recognized scientist of his time, Ostwald's fundamental warnings were ignored. At the start of the 20th century, energy consumption was still comparatively low. The world's population was only 1.5 billion compared to today's 6.5 billion. Electrification was still in its infancy, as was the world of the automobile. Commercial aviation had not yet begun and trade and transport volumes were significantly lower. There were few electrical household goods consuming power, and no radio or television. According to your standpoint, Ostwald's fundamental warnings came either too early or too late. Too early, because the problem had not yet become urgent. Too late, because the fossil-based power industry was already firmly established and exerting a decisive influence on politics, industry and, last but not least, technological developments.

The fossil-based power industry at the start of the 20th century already had a history stretching back 100 years. It had started with a technological revolution; James Watt's steam engine. Patented in 1769, the steam engine initially ran on charcoal before gradually moving to coal. It became the motor of the Industrial Revolution, used in industrial production and later in steam navigation and the steam engines built for the emerging railways. Eventually it was used in steam-driven power plants and is the technology upon which most large power stations are based today, from coal-fired to nuclear power plants. Our modern power industry began with the steam engine, before it developed into a coal industry and subsequently a petroleum and gas industry. It is an energy-burning economy. Only one other technological development in energy is of equal significance; the combustion engine. It was the combustion engine that sparked the automotive revolution and made aviation possible. Aviation, too, shaped and focused its technologies on fossil energy (already the dominant energy source), giving the fossil energy business a further, even greater boost. Seen in this light, a single technology has set the course of the past two centuries, accidentally and with the best of intentions, although with an unexpected chain of consequences.

Pandora's box had been opened. In Greek mythology, it is Pandora, together with Prometheus, who symbolizes mankind's energy drama. Prometheus represents a new energy concept, or the search for such, for it was he who stole fire from Heaven and gave it to mankind. Zeus, father of the gods, saw this action as sacrilege for it brought mankind great misfortune, although, in its enthusiasm, mankind failed to perceive this. Zeus chained Prometheus to a rock as punishment. It was no longer possible to take fire away from mankind, and so Zeus decided that mankind too should be punished; he created the figure of Pandora and gave her a locked box containing all evil temptations. Curiosity having got the better of her, Pandora opened the box and these temptations were scattered throughout the world. All that remained in the box was the hope of a better world. Prometheus is therefore a symbol of the presumptuous striving for opportunities which exceed the scope of mankind, with Pandora as the temptation to carelessly set free the evil which results.

The 1950s heralded the arrival of nuclear energy, putting paid to fears that all of society's energy supplies had become comprehensively, and almost exclusively, reliant on finite fossil reserves. Nuclear energy was praised as a clean, and potentially dirt cheap, alternative to the fossil energy reserves which were drying up. It seemed a new, man-made Promethean gift. In the euphoria which surrounded nuclear energy, all the warnings that society could be diverted from a fossil into a nuclear dead-end were drowned out. Highly complex nuclear power proved mesmerizing. Atomic scientists were treated with the greatest respect. It was unimaginable that nuclear fission, an achievement which far surpassed any earlier physical discovery and which was developed for the construction of atom bombs, could not also be put to outstanding civil use. Early critics of nuclear power such as Karl Bechert, a renowned professor of science and member of the German parliament, were marginalized. Bechert insistently warned of the intractable dangers of nuclear power. In 1957, his was the only vote against Germany's law on the use of atomic energy, the *Atomgesetz*, which paved the way for Germany's peaceful use of atomic power. Even in his own Social Democratic Party, his warnings fell upon deaf ears and his protest was regarded as a nuisance.

The first post-war generation were traumatized by the August 1945 bombing of Hiroshima and Nagasaki with atomic bombs. During the Cold War between East and West that commenced shortly afterwards, this generation lived in acute fear of nuclear war. This also helps explain their hopes for the "peaceful use of atomic power" as enshrined in the *Atomgesetz:* the nuclear arms race and tactics of nuclear deterrence made it necessary to demonstrate a more positive side to the destructive technology of nuclear fusion, and so the movement against atomic bombs became one in support of nuclear power stations. Only later, in the 1970s, did it become clearly and widely understood that this hope was also a dangerous chimera. But the pressing conclusion – that we must focus entirely on renewable energy, the non-fossil alternative to nuclear power – which has only now finally become clear, was then still not drawn. Even for the budding environmental movement, it was still unimaginable that renewable energy alone could meet mankind's energy needs. Right through to the 1990s, most scientists and politicians shied away from speaking up for renewable energy as an equal, even superior and more significant, option for power supply. Advocates for renewable energy themselves bowed down to prevailing opinion, often promoting their projects in an apologetic tone: for they were aware that by doing so, they were heretics in an "atomic age".

A. THE POWER OF THE ESTABLISHED SYSTEM: THE WORLD VIEW OF FOSSIL AND NUCLEAR POWER SUPPLY

For decades, nuclear energy has diverted attention from the fact that renewable energy is the original alternative to fossil energy. Nuclear energy was assigned the leading, and even exclusive, role in supplying energy in the post-fossil fuel age. If, 50 years ago, we had instead focused as intensively on renewable energy, then we would probably not be faced with the current climatic threats to global civilization. We wouldn't have had to engage in resource wars such as the Gulf War or the Iraq War. There would have been significantly less air pollution and disease. Nor would we have the radioactive waste that we don't know where, when or how to permanently and safely store, and the associated problems and costs with which we are saddled for unimaginably long periods. We would probably have had a clean technology industry, few if any environmental refugees, and less poverty in the developing countries. We would be living today in a world free of collective fear of the future. World civilization could be sure of leaving future generations with the same opportunities, instead of heaping unbearable burdens upon them.

These hypothetical scenarios are more than wistful retrospection for, in the 1950s, there really were more technological opportunities available for renewable energy than nuclear energy. Before even a single nuclear power plant had been built, there was already widespread experience in generating renewable energy. It had been understood how to produce electricity from wind ever since a Danish schoolmaster, Poul la Cour, built the first wind power plant in 1891. In the 1930s, several million wind generators were already operating in the agricultural zones of the US.[4] France had already demonstrated how to produce electricity in solar thermal power plants, as evidenced in Marcel Perrot's book, *The Golden Coal*.[5] The technology of producing electricity using photovoltaics also made advances during the 1950s, initially developed for space travel as described in the seminal work by Wolfgang Palz on the history of this technology.[6] That electricity could be produced by water-driven turbines was, in any case, general knowledge: the history of electricity production began with numerous, small hydroelectric plants before the trend to large-scale hydroelectric plants and ever larger reservoirs took off. Decentralized power plants on rivers have always offered great potential, one that has been increasingly ignored. Examples also abounded of the use of biogas, as well as biomass fuels. The technology needed to develop options such as these into reliable power supply systems has always been less complicated and costly than that involved in nuclear energy. However, despite the continued discrimination against renewable energy compared to nuclear energy in national research policies, today the technical and economic opportunities for a transition to renewable energy are so far developed that the complete transformation of our power supply systems can be easily implemented without further delay, allowing us to draw the line under the nuclear and fossil eras.

During the 20th century, fossil and nuclear energy-based power production became the general model for power supply, one involving a fixation with large-scale power plants and the power supply networks constructed on their behalf. General models which are followed overall several generations become axioms, i.e. basic assumptions that no longer need to be proved and the questioning of which is taboo. In the world of science, these models become paradigms, determining schools of thought and ruling out contradiction. This scientific consensus was carried over into politics, business, and society as a whole. It determined decisions which took on physical forms, becoming a matter of course for the majority of the population. People usually try to understand the world in terms of what they see and according to their patterns of behaviour. A paradigm becomes a world view, and this world view determines our thoughts and deeds – even unconsciously – if alternatives are presented. These paradigms are even more persistent where strong interests with the ability to influence opinion are determined to uphold them. The result is a blinkered outlook and limited perception.

The most famous historical example of the obstinate adherence to a particular world view is that of the Catholic Church when faced with the realization that it is the Earth that moves around the sun. Astronomer Nicolaus Copernicus' (1473–1543) discovery contradicted the geocentric world view that had become an article of faith for the Church, one which regarded the Earth as the centre of the universe and mankind as the central figure of creation. Yet the conflict over this discovery only flared up when the new heliocentric world view proposed by Galileo Galilei (1564–1642), the most famous scientist of his time, was confirmed and spread publicly. As a result, Galileo was accused of heresy. It took nearly 360 years before Galileo was publicly rehabilitated, by Pope John Paul II in 1992 (and after an examination lasting 13 years!), at a time when the heliocentric view of the world had already long been publicly acknowledged. The confession that renewable energy offers a universal perspective for supplying mankind's energy needs symbolises a change in thinking about energy and the practice of supplying energy of Copernican-Galilean proportions. The disdain for renewable energy, and the branding of its advocates as heretics, has not endured as long as it did for Galileo but, for more recent world developments, the consequences of this delay are certainly greater.

The advocates of established power supply too, have their curia, their theologians, and a well-established organizational following – power companies, international energy institutes (International Atomic Energy Agency, IAEA; European Atomic Energy Agency, Euratom) and national institutions. For decades these organizations have monopolized discussions about energy, and represent traditional energy thinking right up to the present day. The transition to renewable energy requires a new way of thinking about energy, if only for reasons of physics. No power-generating system (i.e. the overall technological, organizational, financial and political outlay required to make energy available) can adopt a neutral position when it comes to its primary energy source. It would be a blatant mistake to retain the structures that are tailored to fossil and nuclear fuels and simply to exchange the primary energy sources. No technological, organizational, financial or political requirement for producing energy can be seen or understood independently of its primary energy source. We must consider the technological, economic and sociological perspectives in order to comprehend the differences between each primary source of energy, differences which resonate far beyond power supply itself. All we have to choose is the energy source itself, the "gene" of a power system. After selecting the primary energy source, it is this "gene" that indirectly determines what must be done to access and maintain it. We must unfailingly follow the various physical and technical laws of each source of energy, along the entire energy stream, from the point of its appropriation through to its end-users.

The recipients of any form of power supply are the end-users. In the end, consumption is always decentralized, whether in large quantities in a factory, concentrated spatially in a city, or in numerous smaller formats and quantities in the home or car. The decentralized consumption of energy is the only compelling similarity between traditional and renewable power supply. The primary sources of nuclear and fossil energy exist in only a few places on the globe, with concentrated reserves of coal, uranium, petroleum and gas lying under the surface of the Earth. From these locations, these fuels are transported long distances to power stations and refineries and to the billions of energy consumers, to almost everywhere on Earth where people work and live. Thus the location of energy extraction is decoupled from that of energy consumption. Only large energy concerns, operating internationally or cooperating across national borders, can successfully manage this energy stream, from the small number of concentrated reserves in just a few countries to the billions of energy consumers throughout the world. And because no interruptions to this stream can be allowed, the energy concerns are contingent upon close cooperation with governments, and vice-versa. Thus governments have become an integral part of the nuclear and fossil energy industry. Just as we have a military-industrial complex, so too has a political-power industry complex emerged. The traditional power industry has been able to make itself indispensable (and remain indispensable) wherever fossil and nuclear energy has not been replaced by renewable energy. It has enchained society and appropriated for itself the image of "custodian of the economy" – with governments, in turn, becoming "custodians of the power industry". Predetermined by the nature of the energy sources, the power industry has achieved not only the position of a monopoly or oligopoly, but also an intellectual monopoly. It has formed the world view of power supply, not through conspiracy, but through the inherent conditions of the chosen energy sources.

This is also the reason why renewable energy (except where it fits into the centralized power supply system which is structured according to the geographical locations of traditional energy sources, e.g. large hydroelectric plants) was, and is still, universally met with a lack of understanding and reservations about limited levels of efficiency. Although also required to meet the demands of decentralized consumption, for the traditional power industry a system of widespread, decentralized production based on renewable energy is almost inconceivable. The primary source of renewable energy is free, a point made most succinctly by Franz Alt, the journalist who has done most for the cause of renewable energy worldwide, in saying that "the sun doesn't send us a bill". Thus no one can monopolize access to renewable energy. As a natural resource, it is available whether we actively use it or not. Its global potential is unimaginably large. The astrophysicist Klaus Fuhrmann has calculated that, every second, the sun converts 4 million tons of material into energy which it then radiates. This is the equivalent of 386,000,000,000,000,000,000,000 (386 sextillion) watts per second, of which half of one billionth reaches our planet.[7] That is still 20,000 times more energy each day than mankind's current daily energy requirements. It is laughable that anyone could doubt its sufficient potential for meeting human energy needs.

As a natural source of energy emanating from our surroundings, renewables are available globally, although at varying intensities. They enable energy production to be decentralized, thus bringing together the location of power production with that of consumption. Transporting the primary sources of energy is – apart from bio-energy – neither necessary nor possible. We are no longer dealing with physicists' ambitions to constantly increase energy concentrations – from fossil energy to more greatly concentrated nuclear fission, and to the highest level of concentration, represented by nuclear fusion. This focus on concentrating energy is the unavoidable consequence where primary energy sources are acquired and processed by a few centralized power suppliers who then sell and distribute power to the multitudes. Renewable energy allows us to take the opposite path – to enable everyone to appropriate and transform primary energy and, in doing so, to throw off the chains of existential dependency entirely. It is a path away from increasing energy heteronomy to growing energy self-determination, both for individuals and for societies. It leads us away from mankind's separation from natural life cycles to our reintegration, from the crudeness of globalized power supply structures to a multiplicity of structures and a new global division of labour.

For the high energy physicist, this is a step backwards. The same applies for the power industry and energy policy, which has developed from small power plants and municipal supply structures into increasingly large ones. And now all this should be reversed? The only costs involved in renewable energy are technology costs. Its appropriation does not need to be centralized. It doesn't need to be organized "upstream" –to borrow from the language of the petroleum industry – in order to be distributed "downstream" to billions of end-users. The transition to renewable energy demands a new way of thinking about energy, one which envisions a broad-based system of power supply, with its necessary conversion and retrieval technologies, means of utilization and business structures. It is difficult, not only for the special interest groups who are bound to traditional power structures, to understand the technological and social logic of renewable energy and to recognise its true potential.

B. MISJUDGEMENT: THE HERMETIC NATURE OF TRADITIONAL THINKING ABOUT ENERGY

The unconscious or conscious clinging to traditional ways of thinking about energy is the key reason for the many scientific and political misjudgements concerning renewable energy. For example, in 1977, Hans-Karl Schneider, the former director of the Institute of Energy Economics at the University of Cologne, and for a time also head of the German Council of Economic Experts, declared that "solar energy, wind energy, geothermal energy and other 'exotic' energies simply don't offer more than a 5 per cent potential". In 1990, the German Informationskreis KernEnergie (IK), a lobby group financed by electricity suppliers, explained that although "in Denmark in 1988, almost every hundredth kilowatt hour was generated by wind (which represents 0.9 per cent of overall electricity consumption), a similarly intensive use of wind power is not possible in the Federal Republic of Germany because climate conditions are different". In 1993, in an advertisement published in all major newspapers, the German power industry posed the question: "Can Germany phase out the use of nuclear energy? Yes. However, this would lead to an enormous increase in coal combustion and with it emissions of the greenhouse gas CO_2. For, even over the long term, renewable sources of energy, such as the sun, water and wind, will not be able to cover more than 4 per cent of our electricity needs. Can we answer for taking such a step? No."[8]

In June 2005, Angela Merkel, in her position as leader of Germany's conservative party, the Christian Democratic Union (CDU), declared that "increasing the share played by renewable energy in electricity consumption to 20 per cent is hardly realistic". Two years later, now German Chancellor and while holding the European Council presidency, she pushed through an EU decision stipulating that renewable energy should contribute to 20 per cent of the EU's overall energy consumption by 2020. In 2006, and supported by expert opinion, Germany's Environment Minister, Sigmar Gabriel, claimed that by 2025 the share played by renewable energy in power supply could be only 27 per cent at most – roughly the same share as that played by nuclear energy. Three years later, the Social

Democratic Party's manifesto for Germany's 2009 parliamentary elections demanded that the share played by renewable energy in meeting the country's power needs should increase to at least 35 per cent by 2020, and by 2032 to "at least half". Clearly the courage needed to rely on renewable energy had grown significantly within a short space of time.[9]

Scientific forecasts, too, have also consistently fallen short of the mark, even when issued by renewable energy associations. In 1990 the European Wind Energy Association (EWEA) forecast an installed capacity for wind-powered plants in the (then 15) EU member states, of 4,089 megawatts by the year 2000; the actual capacity achieved by that date was already 12,887 megawatts. In 1998 they issued a new forecast: 36,378 megawatts of wind power by 2007; the actual figure reached by this date, however, was already 56,535 megawatts. Even the forecasts of the EU Commission, drawing on research by renowned scientific institutes, lagged far behind actual developments. In 1996 the Commission published a "baseline scenario" and a more optimistic "advanced scenario". The former spoke of a total of 6,799 megawatts of installed capacity for wind-powered plants within the EU 15 by the year 2007 – a margin of error of 732 per cent over actual developments. The latter, suggesting 30,380 megawatts would be generated by wind and solar energy by 2020, indicated a volume which had already long been surpassed – at 73,504 megawatts by 2008. In 1998, the EU Commission issued a further forecast, that of 47,100 megawatts of wind power by 2020 – again, already surpassed in 2008 at 64,173 megawatts. And for solar thermal energy, a forecast of 10,440 megawatts by 2020 – 13 years earlier, in 2007, this volume had already been achieved.

International Energy Agency (IEA) forecasts, too, also regularly lag behind actual developments. In its *World Energy Outlook* for the EU 15, published in 2002, the IEA predicted a wind capacity of 71,000 megawatts by 2030, a figure already achieved in 2009. Similarly, they forecast a capacity of 4,000 megawatts for photovoltaics by 2020, although by 2008 actual levels were 9,331 megawatts. They predicted a global wind power capacity of 100,000 megawatts by 2020, a figure which, at 121,188 megawatts in 2008, had also been long surpassed. The IEA's systematic underestimation of the role of renewable energy is matched by their overestimation of the importance of fossil and nuclear energy: in 2007, when the price of crude oil was around US$100 dollars a barrel, they forecast that by 2030 the oil price would average around US$62. Two years earlier they had predicted an average price of US$30 dollars for the same following 20 years. The IEA is an international governmental institution representing the OECD countries and its "expertise" is relied upon by governments and companies when making investments, and by financial institutions. Their figures also form the basis for many power industry publications. With their incorrect forecasts, they have contributed significantly to disastrous political decisions, to misplaced investments into traditional energy and to the failure to decide in favour of renewable energy. Nevertheless, governments, and the world's leading economies (e.g. the G8 and G20), continue to commission the IEA to carry out new studies.

The estimates and forecasts for the potential use of renewable energy quoted above show just how far wrong recognized energy experts in particular have been – either because other predictions were not welcome, or because these experts were unable to conceive how expanding the use of renewable energy via decentralized installations would be an entirely different process to the investment planning required for large-scale plants. When these same institutions now issue new forecasts predicting higher rates of expansion, and these in turn are presented as the boundaries of renewable energy's true potential, then these new forecasts too will be open to doubt.

Even the projections of the German Renewable Energy Federation (BEE), an active proponent of renewable energy, is more cautious that it might be. They predict a potential share for renewable energy in German power production of 47 per cent by 2020. This represents a three-fold increase within ten years over the 17 per cent share achieved in 2009. This significantly more optimistic figure is, as usual, regarded by many as unrealistic. Yet it is relatively simple to calculate why, by 2020, the share could actually be much higher. Take the introduction of wind power, for example, which covered around 9 per cent of Germany's net power consumption at the end of 2009, with a total installed capacity of 25,777 megawatts being generated by 21,164 individual plants.[10] This represents an average individual plant capacity of 1.2 megawatts. If the only measure taken were to "repower" these plants (to allow a higher permitted capacity and raise average plant capacity to around 2.5 megawatts) then this alone would triple the contribution played by wind power in electricity production, from 9 per cent to 27 per cent. There is no technical reason why this could not be achieved quickly. It would also reduce the cost of wind-generated electricity.

However, over the short term it would be possible to exceed even this target level, for wind power plants are very unevenly distributed throughout Germany. This difference is not primarily the result of varied wind conditions – rather it reflects the differing political criteria for authorizing the installation of wind power plants.

The following table shows the number of installed plants in relation to the geographical area of each German state. The result is extremely revealing: the number of wind power plants ranges from one per 5.6km in Schleswig-Holstein, to one per 183.7km in Bavaria. The contribution made by wind power to net electricity consumption in Germany's federal states ranges from 47 per cent in Saxony-Anhalt, 41 per cent in Mecklenburg West-Pomerania, almost 40 per cent in Schleswig-Holstein and 38 per cent in Brandenburg (where Saxony-Anhalt and Brandenburg are landlocked states), to only 2 per cent in Hessen, and 0.8 per cent in Bavaria and Baden-Württemberg. These differences can only be explained in terms of policy: the states with the lowest wind power plant densities are subject to deliberate political restrictions.

If, over the past years, all the federal states had practised the same authorization procedures as Saxony-Anhalt, with its one wind power plant per 9.1km then

Table 1 Prospective contribution of annual energy yields from wind power plants (WPP) to net electricity consumption

Federal State	Number of WPP as of 31 December 2009	Installed capacity as of 31 December 2009 in MW	Percentage share of net electricity consumption	km²	km² per WPP
Saxony-Anhalt	2,238	3,354.36	47.08	20,445	9.1
Mecklenburg West-Pomerania	1,336	1,497.90	41.29	23,180	17.3
Schleswig-Holstein	2,784	2,858.51	39.82	15,763	5.6
Brandenburg	2,853	4,170.36	38.12	29,470	10.3
Lower Saxony	5,268	6,407.19	22.78	47,618	9.0
Thuringia	559	717.38	11.04	16,172	29.0
Saxony	800	900.92	7.75	18,413	23.0
Rhineland-Palatinate	1,021	1,300.98	7.40	19,853	19.4
North Rhine Westphalia	2,770	2,831.66	3.63	34,088	12.3
Bremen	60	94.60	3.02	400	6.7
Hessen	592	534.06	2.15	21,115	35.6
Saarland	67	82.60	1.6	2,569	38.3
Bavaria	384	467.03	0.83	70,549	183.7
Baden-Württemberg	360	451.78	0.81	35,753	99.3
Hamburg	59	45.68	0.54	755	12.8
Berlin	1	2.00	0.03	892	892.0
Federal Republic of Germany	**21,164**	**25,777.01**	**8.63**	**357,112**	**16.9**

Germany could have had 37,000 wind power plants in 2009 instead of the current 21,164, with an installed capacity (at an average 1.2 megawatts capacity) of at least 44,000 megawatts. The share contributed by wind power plants to net electricity consumption would then be 16 per cent, instead of the current 9 per cent! Furthermore, if the obstacles to development in the lagging states were to be removed over the next ten years, and assuming an average plant capacity of 2.5 megawatts, then as a result (assuming the repowering of all existing wind power plants to 2.5 megawatts) wind power plants would contribute almost 50 per cent of Germany's electricity supply. When we consider the continually growing potential of solar electricity, in addition to electricity generated from biogas, geothermal energy, a growing number of small wind power plants installed either next to or mounted onto buildings (for which a whole series of new system technologies is currently coming onto the market) and, last but not least, the increasing role played by small hydroelectric power plants, then the idea of increasing the contribution played by renewable energy to electricity supply from 16 per cent to 60 per cent within a decade is not a utopian vision but rather an achievable possibility. These measures alone would increase the share played by renewable energy in Germany's overall power consumption from its current 10 per cent to over 40 per cent. If this was matched by a simultaneous increase in energy efficiency of around 30 per cent over a ten-year period, then the share played by renewable energy in contributing to electricity consumption could already grow to over 70 per cent. The complete conversion of all power supply by 2030 would then be easy to achieve. The necessary effort (and the imputed aesthetic imposition of wind power plants) is far less than the imposition made on society when, instead, we continue to rely on nuclear or coal-fired power plants.

The affirmation of expert pessimism

It is not unusual for new technologies to be met with erroneous assessments which later sound unbelievable. They are part and parcel of our political, economic and technological history, and are expressions of the pessimism typical of experts with traditional outlooks. In 1878, Western Union, then the largest US telecommunications company, declared: "The telephone has too many serious limitations for use as a means of communication. This piece of equipment is, by its very nature, of no use to us." In 1895, Lord Kelvin, then president of Britain's Royal Society, explained that no one could construct aeroplanes which were heavier than air. On the matter of adding sound to film, in 1927, Harry M. Warner, a major US film producer, asked: "Who the hell wants to hear actors speak?" Ken Olsen, head of the Digital Equipment Corporation (DEC), one of the first big US computer companies, stated in 1977: "There is no reason for any individual to want to have a computer at home." In 1982, IBM, then the leading global information technology company, decided against buying Microsoft because it wasn't worth the $100 million asking price; IBM was convinced that the future of computers lay in centralized processors. In 1980, while working on a project for the US telecoms giant AT&T, global consultancy McKinsey forecast that there would only be 0.9 million mobile phones in operation in the US by 2000; the actual figure by that date was 109 million. No automotive company had recognized the importance of electric cars until way past 2000, and they have only just started the rush to bring electric cars into serial production as quickly as possible. Errors such as these result from structurally-bound ways of thinking, recognized experts with tunnel vision, and a failure to understand human needs. And, last but not least, they are also the result of underestimating the market dynamics for technologies which do not rely on only a few major buyers, but rather are purchased by multitudes of individual consumers who have recognized their value for themselves.

C. 100 PER CENT SCENARIOS: FROM TECHNICAL POSSIBILITIES TO STRATEGIES

At its founding in 1988, EUROSOLAR declared its aim of heralding in the "Solar Age", in which only renewable energy would be used. Although the organization claimed that this was a "realistic vision" for the 21st century, it was viewed as an over-ambitious dream. A member of the German parliament, who belonged to the Green party and was regarded as one of its thought leaders, expressed his surprise at EUROSOLAR's vision and stated that, according to current knowledge, a contribution of more than 10 per cent would hardly be possible. The 1995 symposium held by EUROSOLAR in Bonn on approaches to meeting all energy needs through renewable energy was regarded as a novelty. However, by this time a whole series of scientific scenarios which demonstrated this possibility in detail had already been worked out. The first "100 per cent scenario" for generating all necessary energy supplies through renewable energy had been drawn up in 1975 in Sweden ("Solar Sweden"). This was followed in 1978 by France which set no target date, in 1980 by the US with a target date of 2100, in 1983 by Western Europe for 2100, and in 1983 by Denmark, with a target date of 2030. Commissioned by the German parliament in 2002, Harry Lehman drew up a scenario for generating 95 per cent of energy supplies through renewable energy by 2050.[11] In 2007, EUROSOLAR published a study showing how, by 2035, the federal state of Hessen could generate all its electricity supplies through renewable energy. However, none of these studies was awarded public recognition, even when, as in 1980 in the US, published by government organizations (in this case the Federal Emergency Management Agency, FEMA) and drawn up with the aid of the Union of Concerned Scientists, an independent scientific organization which counts many Nobel Prize winners among its members. In mainstream energy discussions, such scenarios were taboo. Even a German Greenpeace representative, when I asked in 2006 why his organization didn't refer to these scenarios in its own publications, replied that "we want to be taken seriously". In the meantime, Greenpeace itself is publishing 100 per cent scenarios.

However, it is only recently that such scenarios are being more frequently published and paid a little more attention. In April 2010, a scenario drawn up by consultants McKinsey, on behalf of the European Climate Foundation (ECF), outlined a 100 per cent scenario for renewable energy in Europe by 2050. It concludes that this would generate no more energy costs than the current energy system.[12]

A variety of different 100 per cent scenarios have been presented for Germany recently. In May 2010 the German Advisory Council on the Environment (SRU) presented three options for generating all the country's electricity supplies using renewable energy by the target year 2050.[13] The first option relies entirely on national sources and is seen as the most expensive of the three options because of the lack of potential for energy storage (although only compressed air reservoir and pump water storage mechanisms were considered). The second option involves a German, Danish and Norwegian power network in which Norwegian hydroelectric power plays a key role as a reserve and balance energy. This option requires the current transmission capacity of

1,000 megawatts to be increased to 16,000 megawatts by 2020, and to 46,000 megawatts by 2050. A third option relies on integrating solar electricity from North Africa (see Chapter 3).

In June of the same year, the German Renewable Energy Research Association (FVEE) presented an integrated concept for covering all energy needs solely from renewable energy by 2050.[14] The concept assumes a European-wide power network for delivering electricity. It assumes the complete transition to electrical cars, the use of synthetic fuels generated from renewable energy in shipping and aviation, and heating generated primarily by solar thermal collectors. The FVEE, too, concludes that it will be no more expensive to supply the nation's entire energy needs using renewable energy than it is using the current power supply system. In fact the opposite might be true – "there are potential savings of €730 billion in the costs of generating electricity and heat alone".

In its *Energieziel 2050*, the German Federal Environment Agency also concludes that, in terms of electricity supply, "it is technically possible for Germany, a highly developed industrialised country, with its current lifestyle, consumption and behaviour patterns, to enjoy power supplies generated entirely using renewable energy by 2050".[15] The Agency presents three methods for achieving this goal: a network of regions, large-scale international technologies, and local autonomy. Again, the complete transformation of power supplies is seen as "economically advantageous", involving lower costs than those "with which we and future generations would be faced, should it come to unstoppable climate change". The Agency's focus lies on the network of regions, i.e. exploiting the regional potential of renewable energy. Of the 687 billion kilowatt hours of electricity it is assumed will be needed in 2050, 35 per cent would be generated by photovoltaics, 26 per cent by inland wind power and 26 per cent by wind power at sea, 3.5 per cent by hydroelectric plants, 7 per cent by geothermal energy and the final 2.5 per cent from waste organic matter. They recommend tightening the rules for emissions trading, focusing more strongly on taxing CO_2 emissions, and promoting the integration of renewable energy into both the market and power supply systems.

There are also increasing numbers of 100 per cent scenarios focused at the level of towns and rural districts. Peter Droege's book, *100 Per Cent Renewable Energy*, gives a good overview, and demonstrates such concepts for large cities such as Munich, or for new cities such as Masdar City in Abu Dhabi.[16] BAUM, the German working group for environmentally-friendly management, has also published a book on 100 per cent scenarios which examines initiatives at regional level.[17] All these examples clearly demonstrate that what is being definitively described as possible for individual, even highly industrialized, countries is, in principle, possible everywhere, despite variations in content and consistency. This applies all the more when one considers that almost none of these scenarios and practical concepts have exhausted the current and full potential of the many options offered by renewable energy. To do so would have considerably complicated the calculations.

In 2009, a global 100 per cent scenario was published in the magazine *Scientific American* by Mark Z. Jacobson of Stanford University and Mark A. Delucchi of the University of California. Entitled *Plan for a Sustainable Future*, it lays out a scenario for the complete transition to renewable energy by 2030.[18] This requires around 3.8 million wind power plants, each of 5 megawatts capacity, 490,000 tidal power stations of 1 megawatt capacity, 5,350 geothermal power plants of 100 megawatts, 900 large hydroelectric power plants of 1,300 megawatts (of which 70 per cent already exist), 720,000 wave power plants of 0.75 megawatts, as well as 1.7 billion photovoltaic installations on roofs of 3 kilowatts and 49,000 solar-thermal power plants of 300 megawatts. Global energy requirements for 2030 are assumed to be 16.7 terawatts when generated using traditional energy, but only 11.5 terawatts when derived from renewable energy, as renewable energy is regarded as offering significant efficiency advantages, such as the energy savings involved in electric cars. The costs per kilowatt hour of energy generated using renewables would be lower than that of energy generated using fossil or nuclear energy. Jacobson and Delucchi excluded the use of bio-energy in their scenario, due to fears of the ecological impact this has on agricultural systems and the emissions it generates. They recommend the use of feed-in tariffs as the political instrument for implementing this scenario, as is overwhelmingly used in Germany and around 50 other countries at the moment. The key statement made in this global scenario is that the necessary investment would be around US$100 billion. This sum is compared to global annual expenditure on fuel and electricity, estimated at between US$5.5 billion and US$7.75 billion for 2009. This means the transition to renewable energy is also the "more economical" solution, even when only the direct costs of traditional energy are taken into account, and the external costs, such as damage to the climate, environment and health, are ignored.

The same applies for Greenpeace's study, *energy (r)evolution*, published in June 2010.[19] It assumes annual global energy requirements to 2050 of 13.2 terawatts, of which 94.6 per cent would be generated with renewables. The major share would be generated by wind power (24.7 per cent), followed by solar thermal power plants (20.5 per cent), electricity generated by photovoltaics (15.6 per cent), water power (11.6 per cent), geothermal energy (9.7 per cent), bio-energy (8.1 per cent) and the power of the sea (4.4 per cent). Greenpeace recommends the use of feed-in laws, the flexibility offered by emissions trading, and a stop to subsidies for fossil and nuclear energy.

None of these scenarios should be taken literally, as if they would or could be implemented on a 1:1 basis. Forecasts offering exact percentages, even down to a tenth of a per cent, and for periods stretching several decades in the future, are neither possible nor necessary. No one can predict how costs, and certainly not prices, will develop for each technology over such a long time period, for it is impossible to predict their productivity levels and technological advances and, above all, the potential players and their motives. No one can evaluate investor motives simply in technical or cost terms. And no one can predict the political developments which will facilitate or hinder the transition to renewable energy, or guide it in a decentralized or centralized direction. Scenarios also fail to show how obstacles can be overcome and conflicts between the various courses of action be avoided. In other words, scenarios are not a replacement for setting political targets and engaging in the corresponding action. What is certain, however, is that the role played by each form of renewable energy will be different from anything predicted in a scenario.

Equally, not all operating plants will achieve the capacities which, for reasons of simplicity and predictability, are laid out in each scenario. Therefore a whole series of technical options also remain unaccounted for in all these large-scale scenarios. But most importantly, all these scenarios, whether on a national, European or global scale, ignore the possibilities offered by innumerable, small-scale systems (whether for producing solar electricity, wind electricity, using geothermal energy, or for the combined production of electricity, heat and refrigeration) as well as the potential represented by integrated methods of capturing energy in buildings and equipment and the various methods of storing energy. But because these small-scale systems are, by their very nature, independent, they can be implemented fastest and most widely, and thus they represent the cultural aspect of energy change. It is therefore particularly striking that most of the more recent scenarios which focus at national level still assume an extensive, global supply network (with the exception of the German Federal Environment Agency's network of regions and local autonomy, and one of the three options presented by the SRU). In contrast to these large-scale network scenarios, the 100 per cent plans and initiatives aimed at municipal and regional level are already being actively implemented.

Therefore we have yet to discover how a comprehensive transition to renewable energy, one that meets all our energy needs, will actually be realized. The share played by each technical option, the capacities, the countries and regions in which they are implemented – all this can and will only become apparent once energy change is actually being implemented. The nature of implementation will vary from country to country and from region to region, according to political, geographical, economic and cultural conditions. Therefore, all scenarios are, in their own way, a form of the *Glass Bead Game*. But their value lies in demonstrating the principal technical and economic viability of generating all our energy needs from renewable energy. Thanks to the growing range of available technologies and their increasing efficiency, the practice of using renewable energy can only become more economical and, above all, more versatile.

Thus scenarios such as these are more realistic even than those regularly published by public institutions (e.g. research centres and international energy organizations such as the IEA) for fossil and nuclear energy which assume fossil energy reserves of a magnitude for which there is no empirical evidence. They are more realistic than scenarios which integrate nuclear plants such as fast breeder reactors into their future projections, although to date no reactor of this type is in operation. Or where, as noted above, the IEA recommends the building of new nuclear power plants, without indicating where the necessary quantities of uranium can be sourced or how to guarantee the long-term, secure storage of the enormous quantities of atomic waste they generate. The *Energy Technology Perspectives 2010*, published by the IEA on 1 July 2010, even go so far in their scenario for world power supply in 2050 as to forecast a 19 per cent share for CCS systems (with 3,000 power plants worldwide), although there are extreme doubts as to the political and economic feasibility of this technology. And nuclear fusion, which is being worked on at a cost of countless billions? Nobody knows if it will ever work; the risks involved are left unspoken, and even its advocates say it won't be available until the middle of the 21st century, by which time the transition to renewable energy should have been long completed. Nuclear fusion is the last hope of traditional thinking about energy, the pattern of thought which has led global civilization into a situation which already appears almost hopeless.

Time is inevitably running out for the energies bound up in underground reserves of coal and uranium, oil and gas. This is the moment of truth for the renewable energies available above ground. The potential offered by renewable energy was equally great 100, 1,000, or 10,000 years ago, and will be no greater in 10, 50, 100 or more years. However, although 100 per cent scenarios are a theoretical aid to helping us quickly leave the fossil and nuclear age behind us, they are not a plan, nor a strategy. Even Al Gore's book, *Our Choice,* subtitled *A Plan to Solve the Climate Crisis*, fails to meet the expectations raised by its title. It provides a descriptive overview of all available energy sources, clearly favouring renewable energy, and focuses its recommendations for action on a CO_2 tax and emissions trading certificates.[20] Sketchy outlines like this don't offer practical help in making energy change happen. The question of *how* and *by whom* this change could be implemented is not answered. Yet pushing through a political plan requires the strategic competence to deal with the conflict of competing interests and structures.

D. STRUCTURAL CONFLICT: THE TENSION BETWEEN OPPOSING ENERGY SYSTEMS

The third major disparity between traditional and renewable energy (the first being limited availability versus permanent availability, and the second being emissions versus zero-emissions) is their systemic difference. This third difference is an objective one and therefore it must not be blurred for the subjective reason of avoiding conflict, and nor may it be thoughtlessly ignored. The failure to consider this systemic difference leads to serious errors in strategic thinking.

One such error is to believe that traditional power industry's hold over renewable energy will be broken once renewable energy has become "competitive" or even cheaper to produce. This is, however, a systemic error. The traditional energy system is organized parallel to the stream of energy which it produces. If the key individual element – the power plant – is removed from this energy stream and replaced by electricity produced from renewable energy, then this has immediate effects on the upstream and downstream infrastructure. The primary energy source previously used must find a new customer or it will no longer be in demand. This has effects on the price of the primary energy and on the economic feasibility of its transport infrastructure. The same applies to the downstream sector, especially for the power transmission networks which are tailored to each power plant location. Where a plant is removed from the network (because alternative electricity can be produced elsewhere), then part of the existing transmission network becomes superfluous. Electricity derived from renewable energy is almost never produced at the same location as electricity generated by traditional means; rather it is usually produced in many locations by small generating units.

Thus, when and how a power company replaces its conventional energy offering with renewable energy is not primarily determined by an isolated examination of the costs of electricity production. Power companies make their decisions based on other criteria. This explains why, for example, a power company operating coal-fired power stations who is also active in coal mining (i.e. securing its own fuel supplies) and has a stake in a transmission network is hesitant about renewable energies, for these disrupt the established system. When such power companies do invest in renewable energy, however, then this takes place preferably outside its own framework. The German electricity company E.ON invests in wind power projects in Great Britain rather than in Germany, so as not to disrupt its own established customer base. Its action is a logical consequence of its system, and is copied by other power companies.

In other words, it is not the cost of generating electricity or the cost of fuel for traditional energies which are the sole, or even most important, criteria for the decisions made by a power company. What is decisive is the company's system costs. This applies equally to the fuel sector: petrol, diesel and kerosene are produced in oil refineries. The byproducts generated by this production process are secondary fuels such as those used in the production of lubricants, fertilizers and plastics. Where one of these byproducts can no longer be processed, then it becomes waste. Internal feedback systems explain why established energy concerns are unable to adapt to substitutions by other providers which disrupt their operations. The power companies are prisoners of their own systems. However, they like to present their specific problem as a general one, by glorifying their company's reasoning as economic or social rationality. They see the introduction of renewable energy from their own perspective, but not from that of the wider social interest. Thus any move they make towards renewable energy is limited and designed to have no negative impact on their established system. Renewable energy is thus, for the time being, only a cover or a supplement. The systemic "worst case" for established energy concerns arises when the breakthrough to renewable energy is made by third parties, rapidly and on a wide front, with the energy concerning losing their control over developments. Consequently, they are forced to be active in the field of renewable energy in order not to be left behind, although they will only ever favour approaches that are deemed compatible with their own established systems.

Just how dangerous the transition to renewable energy is for power companies is clear to anyone who observes what happens when renewable energy gains momentum. We can each of us recognize for whom each development brings advantages or disadvantages. It is a transition:

- from imported energy to locally-sourced energy, in countries where energy needs to be imported (most of the world's countries);
- from commercial to non-commercial primary energy, which needs to be neither extracted nor processed and is also free of charge;
- from a transport infrastructure for delivering primary energy (pipelines, ships, trains, tank vehicles), parts of which stretch halfway around the globe, to a primary energy which requires no transport infrastructure;
- from conventional energy storage systems to new means of storing renewable energy which has already been converted into electricity and heat;
- from a few large-scale power plants to innumerable power plants spread over many locations – thus away from a few providers and concentrated capital and to a multitude of providers and widely distributed capital accumulation and value creation;
- from many high-voltage transmission lines emanating from large power plants, to a network structure based on widely distributed production units at regional level;
- from the current power supply industry to the production of technologies needed to harvest, transform and use renewable energy.

The only exception is that of bio-energy, for here the primary energy must be produced, processed and paid for, a process that can be carried out either on a small or a large scale. Here too, however, the delivery streams and processing chain will differ fundamentally from that of fossil energy.

As the location of energy extraction is, by necessity, separate from that of energy consumption, a separation which is global in scale, the traditional energy system is compelled to become a ghetto of major companies which, in order to maintain their own systems, are becoming ever more international. In doing so, they are following the systemic logic of traditional sources of energy, whether in their role as importers or exporters. With the transition to renewable energy, almost all of the elements in this former system become, by degrees, inoperable, as levels of capacity utilization fall. The transition to renewable energy takes place at the expense of the established power industry and its suppliers, for it is the elements of their traditional system which, step-by-step, become uneconomical. There is not even a theoretical point at which their systems can be simultaneously written off: those elements which have already been written off, or are out of date, are accompanied by new investments.

Although objectively possible, from the standpoint of traditional power companies rapid energy transition therefore appears impossible – and so it is, unless they are prepared to destroy their own capital. Therefore they try either to hinder the transition to renewable energy or to draw it out, and certainly to bring the process of change under their own control. Because they themselves are hampered, so they hamper others. These corporations follow their own particular economic rationality – one which is neither industrially, politically nor socially rational. Unless these corporations are able and willing to put themselves through radical root and branch reform, and to accept the immediate, serious losses this involves, then traditional power companies will be the losers in rapid energy change. But which corporate system has ever been able to cope with such change – especially when chained to so many widely distributed system elements? It is therefore not surprising when an electricity company, deciding to go the whole hog, prefers large-scale solar power plants or huge wind parks at sea, and justifies this as the "more economical" option. But economical for whom? These preferences are based on a systemic rationale rather than generally accepted economic reasons. The question of which renewable energy technology (and thus of its sources) is most economic depends on its use and the system specifications of its investor.

Therefore the transition to renewable energy is also inevitably a conflict between two energy systems with different functions. Renewable energy requires different techniques, applications, locations, infrastructures, calculations, industrial priorities, company and ownership structures and, above all, different legal frameworks! Therefore the supporters of traditional power supply, i.e. the current power industry (which is unable to adopt a neutral position towards all sources of energy, as its own system is designed for traditional energy sources) cannot be permitted to set the pace for the transition to renewable energy. Because energy change needs to be rapid, it cannot be made to depend upon those who have an economic interest in slowing it down. After an extremely heated discussion I had about this on television with the chairman of a German power company, he said to me, in a private moment afterward: "Unfortunately, you're right. But if I admit that publicly, then tomorrow I'm out of a job. What would you do, if you were in my shoes?" I could only tell him that I would never be willing to step into his shoes, and that I had no sympathy considering the millions he was being paid for his professional lies.

In practice, the current structural barriers to renewable energy would remain in place even if corporations managed to break away from the traditional world view of power supply. The driving powers for change are, in contrast, those who are least involved with the established power industry. Any strategy that ignores this fact falls short of its target.

E. MOBILIZATION: ENERGY CHANGE AS A COLLECTIVE POLITICAL CHALLENGE

In his book, *Plan B*, Lester Brown, founder of the Worldwatch Institute and current director of the Earth Policy Institute in Washington, demands a political show of strength and speed worthy of a wartime mobilization to effect the transition to renewable energy. He reminds us how, at the beginning of 1942 after the Japanese attack on Pearl Harbor, and Hitler's declaration of war against the US in 1941, US President Franklin D. Roosevelt mobilized the military and began the immediate, mass production of war ships, fighter planes and tanks, declaring that "no one should dare say this is not possible". These actions included an almost three-year ban on the sale of private cars, allowing the automotive industry to concentrate its entire production capacity on manufacturing the vehicles needed for war.[21]

An exceptional show of strength is also required to end the virtual nuclear and fossil war against human civilization's future opportunities, although this is the only analogy with Roosevelt's military mobilization. The current mobilization for energy change demands an entirely different approach to that of Roosevelt, for we are facing entirely different opponents: we need to focus on the production of new technologies, on the complete reorganization of our economic structures, a new culture of production operating within a new framework – one which is free from state-guided interventions in company decisions. Its aim is to override the structural statism of traditional power supply.

However, Roosevelt does exemplify the ideal of purposefully grouping all necessary powers through unconventional methods. He didn't want to be forced to say that "unfortunately we cannot adequately combat Japanese and Hitler Germany's aggression because this demands too much of our existing economic structures". We, too, must be unwilling to allow the strategic mobilization for the transition to renewable energy to be made contingent upon – and compatible with – the vested interests and structures of traditional power supply. We cannot placate the next generation, forced to deal with the catastrophic consequences of fossil and nuclear energy, with the excuse that, although we could have avoided these catastrophes by making the change to renewable energy, we had to take into account opposing interests. That was more important. We apologise for the inconvenience.

Every strategy for energy change demands that obstacles be overcome, obstacles that, however, differ from country to country. Because the sources of naturally occurring renewable energy vary from region to region, there can be no uniform global strategy for energy change. The monocultures of conventional power supply, which strongly resemble one another at international level, will yield to a myriad of renewable energy cultures. For this very reason, the strategic mobilization of renewable energy must be primarily focused at the level of the individual state – not for narrow, nationalist reasons, but because each strategy needs to reflect each nation's natural renewable energy sources, as well as its economic structure and legal system, both of which are closely bound with traditional power supply in so many ways.

Added to this are the various stages of economic development through which countries go. We have developing countries, threshold countries and industrialized countries, countries with a completely structured electricity market and others with only sparse networks. Some countries are energy exporters, other energy importers, some large countries with low population densities, others small but with high population densities. Therefore the overall transition to renewable energy cannot be based on a *single* strategy, applicable to all, although successful ideas such as Germany's Renewable Energy Sources Act (EEG) can be used as a role model for many countries. However, even this is only possible where there are network infrastructures, and only for supplies of electricity or gas. But it is no longer necessary to expand power networks to cover whole countries (something lacking in many developing countries) in order to mobilize the use of renewable energy. Indeed, this would significantly delay implementation. Moreover, we are not dealing with electricity supply and the electricity market alone, but also with the question of heat and fuel supply, market regulation, regional planning, building legislation and tax law and, last but not least, with the question of each country's political ability to act in accordance with its own constitution.

Two basic principles are decisive in every political strategy for mobilising renewable energy:

- The first is the importance of looking beyond the traditional calculations made by the energy industry, calculations which focus only on comparing the current costs of conventional and renewable energy technologies. These generally ignore the major economic costs involved in conventional power supply which are not reflected in energy prices, i.e. the burden of international payments for energy imports as well as damage to health, the environment and climate. Equally unconsidered are the traditional power supply chain's infrastructure costs, over and above those for maintaining the power plant or refinery. The measure of all things is the economic advantages which accrue from using renewable energy, although not enjoyed equally by every player in the market. *Hence political strategies for mobilizing renewable energy must transform these economic advantages into microeconomic incentives*. However, these advantages (and with them each country's own economic room to manoeuvre in effecting energy change) will be lost should renewable energy be imported from other countries where it can be more economically produced. Transformation strategies must therefore be measured against the yardstick of national and regional economies, rather than isolated economic calculations.

- The second principle is the primacy of renewable energy, legitimated by its unequivocal and greater social value, which must guarantee that regulations are established to force traditional energy supplies out of the market as the share played by renewable energy in supplying our energy needs grows. *Official policies must guarantee that the functions of the traditional power supply system are adapted to those of renewable energy.* The established energy industry regards this as an unreasonable demand, for it must be forced to accept a subordinate position while still making the dominant contribution to power supply. No longer can the yardstick be simply how much renewable energy the traditional energy system can cope with. In practical terms, this means refusing to authorize any new investments in traditional power supply mechanisms which involve payback periods of several decades. Only in this way can we prevent the mobilization of renewable energy being constantly and repeatedly thwarted by the functions of the traditional power supply system. For this reason, neither new large coal-fired power stations, nor new nuclear power plants and extended operating lifetimes have a place in a strategy for energy change.

These two principles can only be enforced once regulations have been established. The first principle demolishes the current power industry's frame of reference and scope for action, turning a closed power supply system into a playing field which is open to many innovators. The second is directed at the power industry's interest in preserving its structures, compelling the industry itself to take on a constructive role during the process of energy change. It forces the industry to choose between fighting to maintain its current status (for a further decade or two, or a new generation of large-scale installations) or adapting to entirely new business perspectives (different dimensions and priorities, and extending outside its current core business).

This structural energy conflict is being fought out in the political arena and cannot be seen separately from the battle for public opinion. Every call for energy consensus, with each energy source having or being allotted its "fair" role, creates a quota system and, by definition, a limited role for renewable energy. However, supporters of the traditional energy system and its protectors in political institutions and parties are failing to notice that the transition to renewable energy is developing its own momentum. From a certain point onwards it will no longer be possible for traditional power supply structures or political institutions to hold back the availability of the necessary technologies. At most, it will be able to slow this process down, at least in democratic and free market economies.

This momentum is generated primarily by the technologies which allow renewable energy to be used autonomously, independently of established networks. The most prominent and significant example of this are buildings which can supply their own energy by drawing from their natural surroundings. It wouldn't be the first time that a technological development undermines or overruns established structures and political obstacles. In the world of power supply, only renewable energy can do this. Any energy change strategy must be aware of this potential, for we can never expect every government and political party to simultaneously recognize the signs of the times and decide on energy change irrespective of the energy interests striving to maintain the status quo. That has never been the case, even during the period of coalition government in Germany, when the Social Democratic and Green parties shared power. Governments must be carried to the fight by a democratic public and by the economic players in renewable energy's technological revolution which has begun to unfold. The key political task is to create space for this to happen, by abolishing all arbitrary restrictions on the autonomous use of renewable energy.

The common excuse of those who support the traditional energy system is that rapid energy change is either unfeasible or too risky. In their long-conceded claim to omnicompetence in questions of power supply, the power industry mistakes itself for the economy and society as a whole; what is unfeasible or too risky for the power industry is disqualified in general. Therefore we must all ask ourselves whether the many blockades and brakes hampering and delaying the transition to renewable energy are simply prevarications or justified objections. In the conflict involved with the transition to renewable energy, it is crucially important to make this distinction. As long as refutable statements are accepted as reasonable and cogent, they will prevent many in politics, in society and in the economy from forcibly driving forward the transition to renewable energy, and taking the quickest route to get there.

2

METHODS AND PSYCHOLOGY OF SLOWING DOWN

Paralysis, delays and (un)willing alliances

For decades, the established power industry has used targeted disinformation to explain why renewable energy could not be a viable alternative. Whilst this arsenal of arguments is now largely exhausted, the industry's new mantras against energy change, which aim to delay and justify procrastination, are so subtle that they sound plausible even to some supporters of renewable energy. Nowadays, no one in the established power industry is willing to appear blind to the dangers of traditional energy sources, and thus energy concerns are pouring millions into advertising their role as promoters of renewable energy. They try to create the impression of attempting to realize real opportunities, although this is easily refuted – a quick look at the German Renewable Energy Sources Act (EEG) shows that more than 90 per cent of the investments resulting from this Act have been made by municipal utilities, communities of users or individuals, and not by power companies for whom these investments would have been easier to finance. In 2009 alone, families throughout Germany together invested more in producing solar electricity than the four major German power companies E.ON, RWE, EnBW and Vattenfall combined.

Two of the traditional power industry's delaying tactics are obvious. The first is to talk up alternatives which are presented either as a "bridge" to renewable energy (supposedly available only in the future) or as their equivalents. This approach is targeted at creating a "renaissance" for nuclear power and introducing "more climate-friendly" coal-fired power plants. This tactic helps to fixate global energy discussions on the dangers to the climate of CO_2 emissions, as if there were no other risks and dangers involved in conventional energy supply. The second tactic is to bring large-scale renewable energy projects into play – projects which are time-intensive and predominantly require the participation of major investors. In doing so, the power companies seek to maintain their hegemony by diverting discussion towards fields of action in which they need fear little competition. This tactic even finds favour among supporters of renewable energy where these are insufficiently aware of the connection between time pressures and structural problems. In their relief that energy corporations finally appear ready to focus on renewable energy, they fail to recognize that this is, in truth, an economically-motivated delaying tactic. This tactic is accompanied by playing down the significant problems involved in nuclear and fossil energy, and simultaneously playing up apparent (or partly true, yet recognizably superable) problems which need to be overcome during the transition to renewable energy.

Willingly or unwillingly, those who allow themselves to be deflected from the urgency of energy change by such "bridging tactics" belong to the alliance of procrastinators whose most popular arguments are as follows:

- Change must take place in *global unison*, or at least in agreement with comparable countries. Change would otherwise be impossible to effect, because of the economic damage one would inflict on oneself. Anyway, on a global scale the actions of a single country hardly count. Closely allied to this excuse is the claim, and demand, that the ideal instrument for guaranteeing a sustainable energy supply is *the trading of carbon emissions certificates based on internationally agreed consumption rates*. This is the only way to bring global climate problems under control and simultaneously to significantly improve energy efficiency, which is seen as more important and more economical than the transition to renewable energy. All other political measures are either counterproductive or need to take a backstage role. I examine this subterfuge in the section "organized minimalism" (p43) based on the ideas followed by world climate conferences to date.
- We must continue to rely on *conventional energy bridges* to renewable energy, as renewable energy cannot yet guarantee sufficient energy supply to meet our needs and is not yet able to replace large-scale power plants. Above all, because of the irregularity of wind and solar radiation, renewable energy cannot guarantee to cover the power supply baseload. The most serious problem is the currently unsolved dilemma of storing renewable energy. I evaluate this thesis in the section "brittle bridges" (p53) which examines the claim that nuclear power plants and Carbon Capture and Storage (CCS) power plants are indispensable.
- As long as renewable energy remains more expensive than conventional energy (making its introduction dependent upon state subsidies), all political sponsorship other than research and development represents a *one-sided intervention in the market*. This creates economic distortions and even curtails developments which boost the productivity of renewable energy. Therefore political programmes designed to develop the market in renewable energy are counterproductive. It is in its "own interest" for renewable energy to carve out its own share of "the market". I throw light on this economic construct in the section "market autism" (p71) which deals with the one-sidedness of, and contradictions in, perceptions of the energy market.
- Renewable energy is most productive, and hence most cost-efficient, when *appropriated in locations where the highest solar radiation and most*

favourable wind conditions predominate. Therefore the precondition for their use is the construction of new high-voltage networks (which are anyway pressingly urgent) in order to balance the regional differences in the distribution of natural energy sources. This hypothesis will be dealt with in Chapter 3 which demonstrates the contradiction inherent in strategies which wish to centralize the structures of renewable energy although its natural occurrence is decentralized (p84).

A. ORGANIZED MINIMALISM: THE CONCEPTUAL TRAP OF WORLD CLIMATE CONFERENCES AND EMISSIONS TRADING

Since the mid-1990s, world climate conferences have been regarded as the pivotal focus for introducing energy change, a change which needs to be agreed globally and implemented nationally. As a result, the world public was surprised and appalled by the embarrassing conclusion of the UN Climate Change Conference in Copenhagen in December 2009, for they had not believed its failure possible. The portents had been favourable: a blatantly pressing problem, optimistic government announcements, insistent appeals to non-governmental organizations (NGOs), a huge influx of 65,000 participants and the presence of 120 government heads, making it a "G120" summit. It was the largest political conference in world history. But the debacle that occurred was not so surprising. The climate conference ran according to the same script as had each of its 14 predecessors since 1995: dramatic "now or never" appeals in the run-up to the conference, small-minded and paralysing haggling during the conference leading to pitiable results, the decision to hold a follow-up conference and finally the denunciation of the guilty parties. The only exception, albeit modest, was the Kyoto Protocol which finally came into force in 2005. Admittedly, it was clear from the outset that the Kyoto Protocol could not prevent a further increase in the emission of greenhouse gases, and so it is easy to surmise that only its "toothlessness" permitted its adoption.

If there is a single guilty party responsible for the meagre results at Copenhagen, then it is the idea of a world climate conference itself, an idea based on two highly questionable premises. The first: a global solution in which everyone is allocated similar obligations is essential, because we are dealing with a global problem. The second: climate protection measures should be seen as an economic burden, with the need to negotiate "burden sharing" based on consensus. In the end, this boils down to "all or none", but although convincing in theory, in practice this is illusory. The real problem is the conformity of thought within the emerging "community" of climate diplomats, international environmental NGOs and climate research institutes. To believe there is "no alternative" is an error.

THE PARALYSIS OF CONSENSUS

The fundamental dilemma facing all world climate conferences to date has been that global agreement on rapid and comprehensive initiatives to protect the world's climate must be achieved by consensus. However, there is a fundamental and irreconcilable contradiction between speeding up action and achieving consensus. The more a binding international agreement directly affects the economic and social structures of individual countries, the harder it is to reach consensus. This is always the case in questions of energy, although (as outlined in Chapter 1) in very different ways and according to the particular conditions of each country. Appeals to government responsibility and goodwill cannot outweigh these differences. It is impossible to achieve a truly substantial agreement involving *equal* and *concurrent* obligations, because the circumstances are too unequal. The Kyoto Protocol itself was only agreed because it freed most countries (including China and India) from any obligation to act. It was clear from the outset that the Kyoto II agreement, designed for the period after 2012, could not permit such exemptions if the efforts to reach an agreement on world climate were not ultimately to become a farce. But this just reinforces the fundamental dilemma with which world climate conferences are confronted.

After long and arduous negotiations, the most which can be achieved is consensus on *minimum obligations* which, however, lag far behind actual climate threats. But the Copenhagen conference demonstrated spectacularly that even this minimum target is hardly obtainable (contrary to the pressure to act and all the expectations and announcements). This conference failed even though its negotiation target had been compromised from the outset and already represented a partial capitulation in the face of world climate change. Emissions of greenhouse gases were only to be limited to a heating up of the Earth's atmosphere (currently +0.7°C compared to the heat balance in the Earth's atmosphere from the beginning of the industrial age) of no more than 2°C. In effect, this accepted a further growth in climate hazards, from the current level of 385ppm CO_2 in the atmosphere to 450ppm. To use an analogy, in the year 2000, the UN published its Millennium Development Goals which included halving the number of the world's starving from its 2000 level of 820 million. How would the world's public have reacted if the Millennium Development Goal had been "not to allow the figure of 820 million starving to grow to over 2 billion"? Why have community members allowed the "two-degree target" to become the yardstick by which all is measured, although they themselves constantly quote the famous study by the British scientist Nicholas Stern, which states that advancing climate change creates significantly more economic damage than can be compensated for by economic growth? How can a fatalistic goal be expected to open up new perspectives?

Normally, any compromise is still better than none, because everyone assumes this involves taking on obligations greater than those which previously existed. But it is exactly this opportunity which the favoured "flexible instrument" of carbon emissions trading thwarts – an instrument designed to guarantee and make implementation of the agreement simpler using "market mechanisms".

Minimal obligations and emissions certificates are a dead end

The carbon emissions certificates issued to those signing up to minimum obligations can be traded or used internationally as credits. Those who emit over and above their permitted volumes may purchase "emissions rights" from others who emit less than their allocated maximum. This trade in "emissions rights" – carbon trading – is accompanied by a second instrument, the Clean Development Mechanism (CDM). The CDM allows companies to exceed their permitted emissions limits by making CO_2-reducing investments elsewhere, effectively buying themselves out of their own CO_2 limits.

In order for the market price in carbon trading to be determined by a supply and demand mechanism there needs to be an upper limit – so-called "cap and trade". A country who, thanks to its own initiatives, emits less CO_2, receives financial compensation when other countries fail to meet their own obligations and are forced to purchase carbon emissions certificates. If every country fulfils its obligations, then there is no trade in certificates. However, the opposite also applies: if one country makes a significantly larger reduction in its CO_2 output, then others are only permitted to reduce their own carbon emissions by a much smaller amount. It was clear from the outset that this "zero-sum game" of minimal obligations would lead nowhere and therefore I voted against the German Greenhouse Gas Emission Trading Act (TEHG), which was passed by the German parliament in 2004. It was based on the EU guidelines for implementing the obligations of the Kyoto Protocol within the EU member states. Only one other Member of Parliament, Hans-Josef Fell from the Green party, joined me in voting against this act. Our rejection was met with irritation, even among spokespeople for environmental organizations, for they were hardly able to accuse us of being ignorant of the dangers to the climate. At the World Climate Conference in Bonn in 2001, where environmental organizations such as Greenpeace and the WWF spoke up in favour of implementing such "flexible instruments", EUROSOLAR warned in its campaign "Our air is not for sale", that carbon trading slowed down the transition to emissions-free energy supply rather than speeding it up.

However, advocates of carbon trading praise it as the most effective and important climate protection measure. Indeed, economists, businesses and politicians have all raised their voices to call for the abolition of any other political instrument, such as eco-taxes or the German Renewable Energy Sources Act (EEG). Many environmental organizations justify their support for "flexible instruments" by arguing that they represent the only steps that can be agreed upon at world climate conferences and conferred on all countries. We are encouraged to engage with the idea in order to develop it further for the next step in defining obligations – the Kyoto II agreement for the period after 2012. Economists favour carbon trading, claiming it is a market tool that stimulates the optimal use of investment funds; as investments in climate protection cost less in low-wage countries, the comparatively higher costs associated with investing in industrial countries can be saved, thereby achieving the same effective overall reduction but at a far lower cost.

Sure enough, all these fears have been realized, as demonstrated in the study *Carbon Trading – How it Works and Why it Fails,* published by the Dag Hammarskjöld Foundation.[22] The cap beneath which carbon trading takes place is identical to the painstakingly negotiated – and for climate protection insufficient – minimum compromise as laid out in the Kyoto Protocol (and the subsequent minimum obligation necessary if the next Kyoto II agreement is to be achieved). The flexible instruments of carbon trading effectively turn this minimum into the maximum, with the economic incentive not to overshoot this minimum! Moreover, countries are even warned from exceeding this minimal obligation by means of unilateral initiatives, with the argument that this would damage their economies.

A significant example of this is the report on climate policy (*Klimapolitik zwischen Emissionsvermeidung und Anpassung*) published in January 2010 by the 29 professors of finance and economics who make up the Scientific Committee of the German Federal Ministry of Finance. This Committee concludes that the "uncoordinated actions of individual states" (by which it means initiatives over and above international obligations) should be avoided as they are damaging, not only to the country undertaking the action, but for everyone. "The efforts of individual countries to influence climate policy, by taking on a leading role and setting their own minimum emissions targets, can encourage other countries to slacken their own efforts to protect the climate. As a rule, taking on a pioneer role leads to higher costs in the pioneer country and fails to secure a significant improvement in world climate. Nor do the particular efforts and pioneering initiatives made by individual countries improve the starting position for global agreements on climate. Indeed they can even hinder such agreements being reached. Reducing the remaining advantage of global climate agreements makes reaching such a solution even less likely." For if one country unilaterally reduces its own emissions by "investing in emissions-avoidance technology", then other countries react by "increasing their own emissions". Thus a country is punished for its "early investment in avoidance strategies". The unilateral avoidance of emissions encourages the "circle of beneficiaries and the circle of benefactors" to diverge, creating the "problem of free-riders". Therefore countries should concentrate unilaterally on measures designed to adjust to resulting climate changes (such as flooding and storm damage), for here the circles of "benefactors and beneficiaries" do overlap.

The Scientific Committee's thesis is the perfect example of the logic of flexible instruments: the idea behind world climate conferences results in absurd consequences and hopeless paralysis in global climate policy. It masks all the other problems and dangers associated with traditional forms of energy supply, and fails to recognize that, irrespective of global emissions calculations, all humans have a fundamental interest in reducing the burden of energy emissions that affect them directly. These climate protection instruments do not touch upon the burdens caused by other dangerous byproducts of traditional energy, which are not greenhouse gases and therefore have nothing to do with global climate protection. Even if we were not faced with the problem of CO_2, our health and environment would still be suffering from other harmful substances and the global energy system would by no means be in good order. The problems of exhausting resources as well as the regional and national economic interests of individual countries are also ignored. It is expected that every country, each with its own particular set of social challenges, will march in step in terms of energy supply and according to the stipulations of the agreed global cap on CO_2 emissions. The important thing is that we follow a theoretical model, however unsatisfactory it may be. In the meantime, there is now a flood of "economic" recommendations arguing along the same lines as the scientific committee quoted above. However, as carbon trading is regarded as sacrosanct, it leads to the dogmatic conclusion that any activity which promotes energy change (whether in Germany or elsewhere) must cease, so long as the market for carbon trading certificates remains the yardstick. Many supporters of this "flexible instrument", who simultaneously press for the rapid expansion of renewable energy, are insufficiently aware of the crudeness of this logic.

We have already fallen into the conceptual trap laid by world climate conferences when we judge all energy measures primarily in terms of the extent to which they reduce emissions of CO_2. Other costs and burdens caused by traditional energies are no longer considered, and nor are all the other negative aspects which alone would justify abandoning this instrument. These include the constantly varying price of certificates. This inevitably makes investment uncertain, due to continually changing amortization rates. Carbon trading takes place within the fossil energy supply system and thus serves to preserve existing structures, to prevent renewable energy innovations and to slow down energy change. The role of conventional energy corporations and their influence on government activities remains largely untouched. Instruments for climate protection become vehicles for conserving the status quo of the fossil energy industry.

Governments wishing to auction emissions certificates from 2013 onwards hope to generate treasury revenues. The fees generated by these certificates are effectively a tax on CO_2, although the bureaucracy and costs involved in collecting fees are significantly greater than simple taxation. Governments will not wish to give up future claims to this source of income which may well motivate them to put renewable energy initiatives on hold. Thus governments become indirect business partners with CO_2 emitters. Carbon trading is one of the fastest growing financial markets, making new speculative bubbles likely. Now that we are truly trading in air, it is probably even less possible to control the spurious orders from speculators which are no more than hot air and which led to the 2008 global financial crisis.

Carbon trading takes place in an artificial market of erratic price movements which cannot be relied upon. As "no true market for pollution rights exists", according to economist Elmar Altvater and political scientist Achim Brunnengräber, we must "turn into a commodity something which is actually untradeable". In neo-liberal terms, this is a political device by which explicitly national and international legalized "pollution rights are devised by the state". This presupposes the "setting of artificial caps in order to assure the scarcity of pollution rights as assets".[23]

An invitation to abuse

Even when practised as intended, the flexible instruments have serious shortcomings and are a clear invitation to circumvention and abuse, as evidenced by the scandal uncovered by the EU's police authority, Europol, in January 2010. Europol investigations revealed that over the previous 18 months, companies had cheated EU member states out of €5 billion in proceeds from carbon trading. Certificates were sold repeatedly and sales tax payments avoided. In some EU member states, up to 90 per cent of all trading served only as a tax avoidance measure. In the meantime, the UN too has uncovered abuse amounting to billions. A report published in July 2010 indicated that 22 chemical manufacturers in developing countries have deliberately manipulated the market for certificates.

The energy corporation RWE invested in the construction of a coal-fired power plant in China that was intended to reduce CO_2 output by 460,000 tons. The company was able to offset this saving against the building of coal-fired power plants in Germany which would then emit this additional quantity. The problem is, however, that the Chinese power plant would have been built anyway. CDM was only designed as an instrument for additional investments designed to reduce CO_2 and which have not already been planned. Thus RWE succeeded in reducing the obligations on its German power plant while simultaneously expanding its own business' influence in China, where the additional greenhouse gases would be emitted. This is a case of improving the German CO_2 balance by generating a global increase in CO_2 emissions! RWE employs around 40 members of staff solely for purchasing certificates. This is certainly not the only case of this kind; it is equally certain that other organizations act similarly.[24]

When abuse such as this takes place in countries with a democratic public and a relatively effective public administration, then we can assume that the situation in the many countries with far less transparent administrative structures is much worse. Criticism of these climate protection instruments grows, especially from environmental organizations – even those who had long adhered to the idea of allocating emissions certificates. There are many well-intentioned suggestions for improvement: certificates should no longer be distributed free of cost, at least some of them should be auctioned off; a single CO_2 price should be set; CDM should be more strongly controlled to ensure that it is truly being used to reduce extra CO_2 emissions; all countries, without exception, should be obliged to reduce CO_2 emissions. The most far-reaching suggestion would be, for reasons of global justice, to allocate every single person the right to emit two tons of CO_2 annually, and to calculate each country's obligations according to its population size.

However, no one can explain how consensus on these stricter suggestions could ever be reached, as efforts to secure comparatively harmless obligations have, to date, largely failed. Simply hoping they might be realized just because everyone is willing and ready to follow the theoretical model to the letter fails entirely to reflect socio-political realities. The many suggestions for improvement have neglected to analyse the reasons why the current negotiations have taken their shameful course. Perhaps it is for fear of recognizing that the acclaimed approach was fundamentally wrong – no one wishes to lose face. World climate conferences are caught in the trap of their own fundamental principles, a stillbirth. How many more world climate conferences are necessary before we can admit this? When, and after how many further increases in greenhouse gas emissions, will we be ready to accept government and international organizations' loss of authority? When will we recognize that this loss of authority will only increase and lose us more time if we don't draw a line under these efforts? What has happened to the political realism that is so frequently lauded?

Despite their obvious drawbacks, these "flexible instruments" are continued for ever more and varied reasons. International climate diplomacy, working hand-in-hand with the UN Climate Secretariat and national authorities, has become a self-serving system. It has led to the creation of a climate protection business, consisting of emissions traders, lawyers and well-paid certifying bodies, from which too many already profit. This masks the secret hope of many governments that, for reasons of basic self-interest, they can avoid the obvious solution of changing over to renewable energy. No one can remain unaware that world climate conferences, run by unwilling governments and damned to consensus, are no more than sideshows. Back "at home", governments can speak up for more consistent measures within a framework of measures to be agreed at international level, secure in the knowledge that these measures will be watered down or rejected at international level. This makes it even harder to understand why environmental organizations primarily focus their criticism on agreements either unmade or abused, and less on the idea of agreements itself. Some have even adopted the compromised negotiating target – the "two-degree target" – as their own.

The productive function of independent initiatives

Energy change can only be effected via a revolution in energy technologies. The technological revolutions of our modern industrial history provide a suitable role model. None have depended upon international agreements; many were achieved relatively independently of politics. The political or entrepreneurial drive behind each revolution was to be faster than the competition and to carve out an advantage for one's own economy or company. There is a fundamental reason why this experience is not drawn upon for the revolution in energy technology – governments are willing to protect the traditional power industry and, in turn, the industry openly expects and demands to be protected by its governments. The traditional power industry is the product of political protectionism and has every intention of prolonging its own existence by means of world climate agreements. Warnings against "independent initiatives" are designed to head off changes to the traditional energy system, especially national initiatives aimed at the transition to renewable energy.

We constantly hear the spurious argument that no one benefits when one country forges ahead alone. However, practical developments have long proved the opposite. The German Renewable Energy Sources Act (EEG) has encouraged dozens of countries to promote renewable energy in a similar way. It has sparked a dynamic development of global proportions. It is an independent instrument and has nothing to do with the Kyoto Protocol. It aims not only at protecting the climate but also (as mentioned above) the environment, at minimizing the economic costs of energy supply by incorporating long-term external effects into its calculations, at protecting fossil energy resources and continuing to develop renewable energy technologies. To measure the EEG simply in terms of its current effect on reducing CO_2 emissions (and the costs involved) is to subsequently accuse it of being based on limited assumptions, and evaluating it from a correspondingly blinkered viewpoint. This is typical where attention is focused entirely on the problem of CO_2.

EEG more successful and cheaper than carbon trading

In Germany alone, the EEG has led to a significantly greater drop in CO_2 emissions than officially demanded by the Kyoto Protocol. The Kyoto Protocol set a CO_2 emissions reduction target of 10 million tons by 2012 for fossil fuel-based electricity generation and the industrial sector, whereas the EEG stimulated investments in renewable energies which, by 2009, had already reduced CO_2 emissions by 66 million tons. In Germany, emissions certificates had led to an increase in electricity costs of €10.75 billion between 2005 and 2007 alone. Although the certificates were issued to power companies free of charge, these same companies include the market value of these certificates when calculating their customers' electricity bills. In all, as a result of carbon trading and the largely free distribution of CO_2 certificates, these windfall profits for the major German power companies will increase to around €44 billion by 2012. In contrast, the extra costs for introducing renewable energy will be just under €47 billion for the period 2005–2012, and will have led to a reduction in CO_2 emissions in Germany of over 80 million tons, representing an eight-fold monetary effect.[25] However, whereas promoting renewable energy leads to generating of environmentally-friendly electricity, technological developments and cost degression for future technologies (quite apart from avoiding imports of fossil energy and other external costs), the power companies' windfall profits do not reflect any positive contribution. These are all concrete figures, in contrast to the Kyoto scholars' theoretical constructs of the costs of reducing CO_2.

The EEG upon which the power companies focus their attacks has probably indirectly encouraged more climate protection measures worldwide than the entire Kyoto Protocol. It has sparked the mass production of renewable energy technologies, reducing their costs and building up production facilities for these technologies in countries such as the US, China, India and Japan. In turn, these countries produce and sell these technologies to the rapidly growing German market, as well as other markets with laws similar to the EEG. This is how technological revolutions take off – by igniting a self-supporting dynamic process. Those who prefer certified quotas, distributed according to one-sided and questionable efficiency criteria, end up with a technocratic state-directed economy rather than a technological energy revolution.

Even so, the chief ideologues of carbon trading (with either a short-sighted or deliberately blinkered viewpoint and an audibly two-faced portrayal of the facts) bemoan the "economic inefficiency" of the EEG and the "highly efficient nature of carbon trading". Particularly frequent and vocal are the lamentations of the Rheinisch-Westfälische Institut für Wirtschaftsforschung (RWI), a scientific research and policy advice institution based in Essen, which measures (as if no other criteria existed) every possible cost associated with reducing CO_2 emissions, before misrepresenting them. According to the RWI, it costs €900 to reduce emissions of CO_2 by one ton using solar electricity, €200 per ton using wind generated electricity, and only €30 per ton or less as the result of carbon trading. How do they manage to come up with such figures, which are starkly disproportionate to the prices customers are charged for electricity generated using renewable energy and for electricity that factors in the price of CO_2 certificates? Here we see two incomparable and opposing ideas in collision: the RWI (taken here as a symbol for all advocates of carbon trading) compares real investment costs in renewable energy systems (and converted into the zero-emission electricity these systems produce) with the trading price for allocated CO_2 certificates. It also compares the costs for completely new investments into renewable energy systems with partial, additional investments into existing conventional systems in order to reduce CO_2 emissions by a few percentage points. And thirdly, it compares (when referring to new and more efficient fossil fuel power plants that emit perhaps 10 per cent less CO_2)

renewable energy systems which produce zero-emission electricity throughout their operational lifespan, with fossil fuel power plants which continue to emit substantial quantities of CO_2 over a period of 50 years or more. Thus they compare the permanent effect of entirely avoiding CO_2 emissions with the short-term effect of reducing emissions by a few percentage points. In short, even under close economic analysis, they continue to compare like with unlike.

Let's simply look 20 years ahead, at the investments made as a result of the EEG, when the guaranteed subsidies for each system have run out and the systems themselves have long been amortized. By this time, the costs of producing electricity using solar and wind generating systems are perhaps around €0.01 per kilowatt hour or less – the only costs are those of replacing power converters and regular maintenance. In contrast, a power plant run on fossil fuels, even with reduced CO_2 emissions, will also still be operating after 20 years and consuming increasingly expensive fuel, and it will continue to emit CO_2. This is quite apart from all the other emissions and their associated social costs, ones that, for the chief ideologues of carbon trading, no longer exist, as if they lived on a different planet. Thus someone like Christoph Schmidt, head of the RWI and, at the same time, one of the five members of the German government's Council of Economic Experts who are tasked with examining overall economic policy (the so-called "five economic sages"), can come to the ridiculous conclusion that the environmental effects of the EEG are "practically zero" and the overall economic effect is "zero at best". "What is saved as a result of the EEG is simply emitted elsewhere, with emissions being simply transferred to other industrial sectors which are involved in carbon trading."[26]

In an article for the German Advisory Council on the Environment (SRU), economics professor Karin Holm-Müller questions the advisability of comparing like with unlike. It is "self-evident" that this leads to a "preference for coal-fired power plants" although a truly economic "comparison of all the costs" indicates that these are more costly than avoiding CO_2 emissions by alternative methods (e.g. wind power). As a result of this original misjudgement, we continue to adhere to the "long-term socially sub-optimal path of avoidance", i.e. carbon trading. This problem is accentuated by the fact that obligations to reduce ever larger quantities of CO_2 are only ever introduced in stages, from one "trading period" to the next, and always involving arguments, with the magnitude of each stage needing to be individually negotiated. As a result, businesses always prefer to raise only the additional "costs of reducing emissions within existing systems" over investments into renewable energy, because of the shorter terms and more manageable calculations involved. Consequently, "in this way the status quo of established technologies continues to be preserved, enjoying a level of protection which cannot be provided by carbon trading alone". This leads to "new technologies being disadvantaged by carbon trading systems".[27] Thus the waiting and delaying tactics pay off.

B. BRITTLE BRIDGES: NUCLEAR ENERGY AND CCS POWER PLANTS AT ANY PRICE?

Many environmental organizations would have quickly stopped rejoicing over a success in Copenhagen. The world climate change conference in Copenhagen would have recognized two controversial measures as being integral components of official climate protection policy – the construction of new nuclear power plants and the permanent disposal of the CO_2 emitted by coal-fired power plants, either underground or on the seabed. At the beginning of the new century, with the Kyoto Protocol still in the starting blocks, new nuclear power plants were not recognized as instruments for climate protection and there was certainly no talk of Carbon Capture and Storage (CCS) power plants. However, since 2004 there have been calls for a "renaissance of nuclear power", and a little later the concept of CCS power plants was introduced. To avoid the question of why new investments should not instead be directed entirely towards renewable energy, nuclear power and CCS power plants were described as "bridges to renewable energy". Sadly, it is not yet possible to meet energy demands using renewable energy alone, was the habitually repeated justification, a claim which obscured the actual intention which was to continually extend these "bridges", for the next 50 years or more.

The new catchphrase is "carbon dioxide free" energy sources. At the UN Climate Change Conference in Bali in December 2007, terms such as "clean energy technologies" and "zero carbon economy" were added to conference papers and accepted unchecked. They stem from the store of semantics belonging to those who wish to promote nuclear power and CCS power plants into the climate protection ranks of renewable energy. One of the few people in Bali who spoke out against this was the former chair of the World Future Council, Bianca Jagger. She pointed out that only renewable energy could be described as "clean energy". By the Bali conference at the latest, it was clear that a new climate agreement (Kyoto II), due to be agreed two years later in Copenhagen, would officially recognize CCS power plants and new nuclear power plants as climate protection measures. The Intergovernmental Panel on Climate Change (IPCC), the UN Climate Secretariat and most renowned climate research institutes include both in their lists of climate protection approaches. Advocates of nuclear power and CCS power plants have taken advantage of global discussions on energy being fixated on CO_2 emissions, for this serves to push all other dangers associated with conventional energy supply into the background. Therefore, if an agreement had been reached in Copenhagen, many environmental organizations would have been faced with an awkward problem when celebrating as they had done eight years earlier after agreement of Kyoto I: not only would they have had to tolerate increased levels of greenhouse gases, up to the "two degree" cap, but also, and in the same breath, to take a stand against the contents of the agreement, in order to remain true to their unanimous rejection of nuclear power and overwhelming rejection of CCS power plants.

What bridges?

There was never any doubt that bridges would be necessary before all energy supplies could be generated by renewable energy. An immediate jump from conventional energy supply to renewable energy is unimaginable – although neither the technical nor economic reasons are insurmountable. The true reasons are political and social, for it cannot be expected that governments and parliaments will all simultaneously muster the will and strength to extract themselves from the habitual, and occasionally extortionary, influence of powerful traditional energy interests. Additionally, most lack the strategic competence to introduce renewable energy, for this is a subject matter relatively new to governments, political parties and financial institutions. A lack of public education means that popular support for renewable energy is still also very varied, and there is an almost universal lack of trained scientific and technical personnel which reflects the failures in the educational and scientific policies of the past decades. Thus the learning curves in all these areas need to rise sharply, and this in turn demands an across-the-board strategy.

Every single module in a system of renewable energy can be installed and start operating quickly once administrative constraints have been cleared away at political level – energy autonomous buildings, solar power systems, wind generation systems, biogas systems, small hydroelectric plants, etc. Installation times range from a single day to several months. Compare this to any large-scale plants with construction times of several years, nuclear power plants requiring up to a decade and more, quite apart from the time needed to build new transmission grids.

So what are the bridges that lead to renewable energy? One is the existing capacity offered by the traditional system of energy supply. Output from renewable energy systems will grow and gradually supplant that of conventional capacities until the latter become entirely dispensable. Rapid energy transition requires only that operating lifespans for traditional methods of generating energy are guaranteed to be limited to the actual need for their production capacity. Thus as renewable energy is mobilized, traditional energies are simultaneously forced out of the market, and governments must relinquish their role as protectors of the traditional power industry. This is of course anathema to established power industry players, but it is the normal process in any technological revolution. Take the mass introduction of PCs, for example. Who paid attention to the interests of typewriter manufacturers who are now all but gone?

The bridges to renewable energy must be short and accompanied by strategies which smooth out the path of energy change and help drive it forward. These strategies include all means of saving energy and increasing energy efficiency, particularly in buildings, in engines and equipment, as well as avoiding long transport routes with their inherent energy losses. Such measures will make energy change easier and cheaper by reducing overall energy needs. Transitional aids include systems which, although they run on fossil fuels, have no long-term capital requirements and fit into a future system of modular energy supply, such as combined heat and power units (engine-driven power plants with manageable operating lifespans of around 12 years). As well as their efficiency (they are able to use the deployed energy in multiple ways), they also have the advantage of being able to run on a variety of fuels, and can thus increase the share played by renewable energy until fossil fuel can be avoided completely.

In contrast, nuclear power plants and large power plants driven by fossil fuel, whether new or with extended operating lifespans, are not bridges to renewable energy. Indeed they act as absolute barriers, thanks to their long construction times, operating lifespans of between 40 and 60 years, their low levels of efficiency and fuel inflexibility, and especially where extensive new transmission infrastructures need to be installed. Here we're talking about steam power plants which involve unavoidably large energy losses. Steam must be permanently available so that the power plant is ready to drive the turbines which produce electricity at all times, despite constantly varying levels of demand.

These steam power plants are regarded as essential *baseload power plants* because they can produce electricity around the clock, i.e. theoretically 8,760 hours a year. They can also be throttled back when demand is low. Normally, a coal-fired power plant runs at an operating capacity of 95 per cent throughout the year, i.e. for over 8,000 hours. In contrast, because of constant minor or major operational disruptions, a nuclear power plant rarely achieves levels over 70 per cent. A steam power plant requires a warming-up period of around eight hours before it is initially fired up and ready to go, and therefore it makes no sense to close down these plants entirely. But there are benefits and drawbacks to continual operation. One disadvantage is that, even when throttled back, for periods of several hours at a time more electricity is produced than is actually needed. An advantage of steam power plants is that they are almost always able to meet demand. However, as it is not worth using the steam process to meet every peak in demand, gas-fired power plants and pump storage power plants stand by, ready to step in and make up demand, but operating only for a few hundred hours per year.

The more wind and solar electricity that enters the network, the more these baseload plants need to be throttled back, immediately reducing their capacity utilization levels. In order not to become totally superfluous, they need to act as standard and reserve power plants – a role for which they were not designed. As a result, they effectively serve as a technical competitor to wind and solar electricity, leading the SRU to conclude in its statement published in May 2009, entitled *Setting the Course for a Sustainable Electricity System,* that "in a supply strategy based on coal-fired power plants and nuclear power plants, the share played by renewable energy sources must be significantly limited if these baseload plants are to be run economically".[28] The deal offered to the British government by the two energy corporations E.ON and EDF is thus, from their own standpoint, logically consistent: these corporations were prepared to build nuclear power plants in England if the government would agree to cap renewable energy's share in electricity production to 35 per cent.

However, we would never have needed such large power plants if, as Gottfried Rössle convincingly argues, the power industry had developed with a diverse and decentralized structure based on engine-driven power plants.[29] Such plants can be started up as quickly as a car when electricity is needed, can be shifted up or down a gear as required, and also generate heat. All this leads us to conclude that large-scale power plants cannot conceivably be regarded as bridges to renewable energy. In terms of energy technologies and efficiency criteria, nuclear power and coal-fired power plants are considerably more complicated to manage than renewable energy. In terms of climate protection and energy security they are unnecessary, and the social costs are unacceptable. In reality, power companies use nuclear power and coal-fired power plants only to protect and extend the life of the traditional energy supply systems. In several important countries, nuclear power is also favoured for a very different reason – to enable the production, or option of producing, atomic weapons.

Unacceptable nuclear power

At this point we must once more lay out in detail the arguments against nuclear power. They have been stated many times. Since my time as a systems analyst at the German Nuclear Research Centre in Karlsruhe (1976–1980), my attitude to nuclear power has stiffened: even if nuclear power cost nothing, it should still be rejected.

One of the key reasons, as Christine and Ernst-Ulrich von Weizsäcker so succinctly pointed out, is that nuclear reactors lack the "error-friendliness" so essential in any technology. Nuclear power plants have the potential to make errors which have catastrophic and irreversible consequences for entire societies. The worst-case scenario for nuclear power plants is a meltdown of the reactor's core. The 1986 catastrophe at the reactor in Chernobyl is proof that such errors can happen. Several similar catastrophes have been prevented at the last minute – in 1979 at the nuclear reactor on Three Mile Island in the USA, or in 2006 at the Forsmark reactor in Sweden, a country known for its particularly high security standards. The Chernobyl disaster occurred in a relatively under-populated area. If a similar catastrophe occurred in a densely populated and economically important area (e.g. at one of the two Biblis reactors in the Germany's Rhine-Main region, one of the two reactors in Neckarwestheim near Stuttgart, or in one of the two Isar reactors close to Munich) then this would be a death blow to the entire national economy. Catastrophes can also be caused by external influences, such as a direct attack with a hijacked aeroplane flown directly into an atomic reactor. The Biblis reactor in the Rhine-Main area, for example, is only 40 flight seconds from the approach path to Frankfurt's main airport.[30] That a very real danger of nuclear terrorism exists was highlighted at the international conference hosted by US President Obama in April

2010 to which 40 governments were invited. Even without the danger of a nuclear terrorist attack, nuclear power plants have no claim to error-friendliness or immunity to catastrophe. No humans or technologies can always function without error, especially in a reactor in which tens of thousands of technical components, many of which are highly sensitive, need to work in perfect synchronization.

In plain English, because what could happen cannot be allowed to happen, we cannot use technologies which have consequences far exceeding any level of social or economic acceptability. The same applies to the problem of atomic waste which only stops being hazardous after 100,000 years. Which political system will be around that long? And which company currently operating nuclear power plants can secure the permanent disposal of its own atomic waste over such a period of time? Nuclear power is therefore the most foolhardy project in the history of civilization.

The great future promised by nuclear energy – that of nuclear fusion – can play no role in the energy dilemma which needs to be solved in the next few decades because the technology is not yet available. There is no talk of its risks (which, although not identical to those of nuclear fission, are significant) and nor of the costs, which are highly likely to be substantially greater than those for the transition to renewable energy. Hence there will be no need for nuclear fusion. Even the announcement buzzing through the world's media during the spring of 2010, that Microsoft founder Bill Gates wanted to develop mini atomic reactors, fails to save the cause of nuclear energy. According to Gates' plan, these mini reactors should each have a capacity of between 10 and 300 megawatts, be operated completely automatically and require little uranium. But according to the development teams, although there would be more time for taking counter-measures, the possibility of a meltdown couldn't be ruled out. We would still have atomic waste and no details can be given of the time or costs needed for its successful development.[31] The fact that Bill Gates and his team are fascinated by this scientifically and technologically demanding project cannot be reason enough to rehabilitate nuclear energy.

The physicists in the German Physical Society's (DPG) working group on energy, who in March 2010 assiduously demanded the construction of new nuclear power plants, are also unable to let go of their dream of nuclear energy, even though it has long since become a nightmare. How can we allow so much intellectual investment to go to waste?[32] Yet considering the overwhelming potential offered by renewable energy and its zero level of risk, there is no scientifically tenable reason to require nuclear energy.

The nuclear community's drive for self-preservation

The interesting political question is why, despite all the unmistakable risks and the intellectual ambition of atomic scientists, there continue to be attempts to create a renaissance in nuclear energy. The primary driver of these efforts is the self-preservation interests of power companies, the atomic industry, international and national atomic energy institutions and atomic research centres. Construction of nuclear power plants, most of which were built between the 1960s to the 1980s, largely stagnated from the middle of the 1980s onwards. The primary reason was growing civil protest, although this was not universal and is therefore not the only explanation. An additional factor was the shock effect of the Chernobyl catastrophe which frightened many governments, discouraging them from future projects and generating unexpected rises in costs as a result of increased security regulations. Atomic experts had underestimated the complexity of nuclear reactors, as demonstrated by the many accidents occurring in all reactors. The nuclear technology industry has had its hands full dealing with the additional security requirements. Yet now the time has come to shut down many of the older reactors which have come to the end of their operating lifespans. However, every reactor no longer in operation means fewer contracts for the nuclear technologies industry.

The pressure to which the nuclear technology industry is being subjected is clear: as of August 2009, "only" 52 nuclear power plants are under construction and 435 are in operation worldwide. In 20 of the 32 countries with nuclear power plants, average plant age is between 25 and 35 years, so these plants will need to be decommissioned within the next decade. However, in August 2009, only 90 new reactors were on the drawing board. As a result, unless operating lifespans are lengthened and the number of new nuclear power plants under construction increases, within the next decade the number of nuclear power plants worldwide will halve. In their World Nuclear Status Report 2009, Mycle Schneider and his co-authors examined the action needed simply to maintain the number and installed capacity of nuclear power plants at its current scale: in addition to the 52 new reactors currently under construction, an additional 42 need to be completed by 2015. This means that, on average, a new reactor must commence operations every six weeks. In the years to 2025, a further 192 new reactors must come into operation, representing an average of one every 19 days.[33]

The Status Report argues that this is practically impossible, due to the lack of industrial capacity and qualified personnel. Thus French president Sarkozy is feverishly seeking to generate new construction contracts for AREVA, the French nuclear technologies giant, from Italy to North Africa and Asia. At the same time he is demanding subsidies for the construction of new nuclear power plants in the EU as well as massive efforts to train a new generation of nuclear technicians. In March 2010 he organized an international conference on nuclear power in Paris to which many newly industrialized countries were invited. He repeatedly demanded that the construction of nuclear power plants be declared an international development project. These attempts to usher in a renaissance for nuclear energy can only be explained in terms of a "renewable energy lie", i.e. by denying the feasibility of 100 per cent renewable energy. This explains why the Swedish Energy Minister, Andreas Carren, reacted to protests against his government's announcement that they would be building new nuclear power plants by saying, "protests against nuclear power plants are okay, but someone needs to tell us what can be done to avoid emitting environmentally unfriendly substances". If AREVA stumbles due to a lack of orders, then France – which relies on nuclear power plants **and** its atomic arsenal – has a severe problem. If it wishes to hold onto both then it may be forced to provide AREVA with billions in state funding. It would then be clear that the success story of French atomic policy was over, but a bankrupt AREVA would be a political meltdown for France as a nuclear power. In fact, Sarkozy pledged a €1 billion investment in nuclear power in June 2011, at a time when many other companies were turning away from nuclear power.

The existential dilemma of highly qualified atomic physicists

The demand for nuclear technicians creates a massive problem. They are needed not only to operate existing nuclear power plants as safely as possible, but also to dismantle decommissioned nuclear power plants, to prevent radioactive material from being used to build atomic weapons, and to securely store and monitor atomic waste for an unimaginably long period of time.

These tasks require highly qualified and responsible personnel. However, the moratorium on building nuclear power plants over the last 20 years or so, and the increasing controversy surrounding nuclear power (leading to referendums against nuclear power in Austria, Italy and Sweden, and Germany's decision to phase out nuclear power) has meant that up-coming scientists have failed to see any career prospects in nuclear energy. Today's nuclear physicists and technicians are aging.

This acute personnel problem highlights the profound dilemma posed by nuclear energy: how can we ensure that there are sufficient numbers of highly qualified nuclear technicians, over the mid and long term (even after the necessary and unavoidable end to using nuclear power), who will be happy to act as gravediggers and cemetery guards for millions of tons of stored atomic waste? Who wants to train for such a career? This is an almost insoluble problem, one which the pioneers of nuclear power did not consider, having imagined we would use nuclear power for eternity. We can manage without nuclear energy, but its legacy can hardly be undone. The plan for escaping this looming personnel emergency is to prevent it happening, by perpetuating the use of nuclear energy. Because we have started, we have to go on, in perpetuity. In a way, society has been taken hostage. Nuclear energy operators are also captives, and this determines their standpoint and behaviour.

The nuclear industry's payroll remains large and influential although it has lost its role as the symbol of our future (a role it was awarded in the 1970s). In 1974, the International Atomic Energy Agency (IAEA) was still talking of over 4 million megawatts of installed capacity by the year 2000, a sum far greater than current global capacity. Yet the national and international public institutions (national research centres and the international organization Euratom and the IAEA) which were founded in the 1950s remain largely untouched and promote the use of nuclear power for generating all future power supplies. The IAEA has over 140 member states and more than 2,000 employees. When founded in 1957, its intended role was to direct future global energy supply. It will come as no surprise that, together with Euratom, the IEA and their sub agency, the Nuclear Energy Agency (NEA), national atomic energy institutions, the nuclear technology industry and the nuclear power companies, they are not prepared to accept that their role is simply to wind down the use of nuclear power.

The self-confidence and influence of this nuclear body with its strong international networks is too great not to make a new push for nuclear energy, and threats to the climate provide a welcome opportunity for doing so. Proponents of nuclear energy didn't begin their specialist careers simply to function as industry liquidators. It must be unbearable for them simply to work for an outdated model. Therefore, in order to carve out a future for nuclear power, they must publicly adhere to the thesis that renewable energy does not represent a sufficient and secure alternative, even when this statement has no scientific basis and has long since become laughable. Defining nuclear power as a bridging technology for use on the path to renewable energy is thus a tactical admission. Making nuclear energy generally more popular requires massive efforts at public promotion, hence the artificial notion of a "nuclear renaissance", with its promises of secure nuclear power plants and large-scale expansion plans as propagated, above all, by the IEA. They know this cannot be realized, even if governments and populations all strove to achieve this goal and were free from financial constraints. But they hope that, by drumming up support for nuclear energy, they can at least preserve their own role.

A bogus proof of progress

The nuclear plans and recommendations being discussed also affect developing countries. Here they are met with considerable acceptance as nuclear power continues to be regarded as proof of technological advancement. This is a belief fed by the IAEA, which uses its large presence and global communications network to run hundreds of workshops for scientists throughout the world, encouraging them to agree on the uniqueness of nuclear power. During a speech on renewable energy in Vietnam (with a nuclear energy commission reporting directly to the head of government although it currently has no nuclear power plant), I spoke with some of the country's most respected physics professors. They were all familiar with nuclear power, none saw any fundamental problem with it, and their state of knowledge about renewable energy was at a level typical of the 1970s. My experiences to date in many other countries have been similar, from Latin America to Africa. At a conference of the Royal Academy of Science in Jordan, I argued against the suggestions of one of the country's most well-known energy scientists who recommended the building of nuclear power plants. When I asked how he could put forward such a suggestion, considering the enormous volumes of water required by nuclear power plants (3.2 litres of water for each kilowatt hour of electricity generated by nuclear power) and knowing that his country suffers an acute water crisis, many in the audience were surprised and he was unable to answer.

Because of the work of the IAEA (mandated and showered with the necessary resources), 23 countries with no nuclear power plants for generating electricity already have nuclear reactors for research purposes. This includes countries such as Egypt, Morocco and Libya, who are particularly rich in solar radiation or wind power, and countries such as Algeria and Georgia, Indonesia, the Philippines and Thailand which, in addition to their wealth of naturally occurring renewable energy, are beset by internal unrest. The IAEA regards these countries as potential newcomer states for nuclear power plants, as well as 15 other states including Bosnia, Uganda, Jordan, Namibia, Nigeria and Tunisia. According to the World Status Report on Nuclear Energy, most new reactors are being built in China (16), Russia (9), India (6), and South Korea (5). Most of the concrete plans for new reactors are concentrated on China (29), Japan (13), the USA (11), India (10), Russia and South Korea (7 each). Construction times average five years, although delays regularly occur.

All this is taking place although no reliable permanent disposal for nuclear waste is in sight, and exploding costs for building new nuclear power plants are disproving the claim that nuclear power brings cost savings. Citigroup Global Markets, which operates worldwide, warned the British government in an appeal titled *New Nuclear – The Economics Say No* against plans to build ten new nuclear power plants.[34] As the government was not offering to provide any financial support for their construction, they argued that companies would be faced with unacceptable risks in the form of unforeseeable rises in costs. They pointed to the exploding costs of constructing the new Finnish reactor. The fixed price had been given as €3.5 billion, and with construction being completed between 2005 and 2009. Yet the costs have already risen to €5.5 billion and construction will be completed by 2012 at the earliest. By that date the costs will have risen again. Citigroup also names other examples of new reactors whose construction costs are double or more than original estimates. Consequently, we must assume a corresponding increase in the price of generating electricity, rendering these nuclear power plants no longer competitive.

However, if new reactors are to be built, then this is only because energy corporations assume that state aid will be forthcoming. France's President Sarkozy is demanding exactly that – a new wave of nuclear power subsidies, either for building reactors or for covering actual production costs, with the price for electricity generated using nuclear power being set by politics and with purchase guarantees (along the lines of the EEG's guaranteed feed-in tariffs for electricity produced from renewable energy). The demands of nuclear power plant operators are clear – "equal rights for nuclear power". But there is a key difference. Nuclear power has been subsidized to the tune of countless billions over the past 50 years, and it is becoming ever more expensive. In contrast, renewable energy has only been recently introduced and has received significantly less public subsidy to enable mass production of renewable energy systems. Quite apart from other qualitative differences, such as the social value of renewable energy compared to the social costs of nuclear power, renewable energy is also becoming increasingly cheaper.

CCS – asphyxiating gases for politics and society

Carbon Capture and Storage (CCS) power plants are power plants in which the CO_2 emitted during the coal or gas combustion process is captured and stored in underground repositories or on the seabed. CCS power plants are therefore coal-fired power plants with chemical factories attached, as well as a CO_2 pipeline infrastructure and CO_2 permanent disposal technologies. As a consequence, the cost of generating electricity is far higher than in today's fossil fuel power plants. Even so, despite the unclear costs and the completely incalculable environmental risks, the use of such "climate-friendly coal-fired power plants" has already become a fixed component of national and international climate protection strategies.

Those who currently favour, and are driving forward, this CCS option accept that, by doing so, the substitution of coal-fired power plants by systems for generating power through renewable energy will really be delayed until the second half of the 21st century. If CCS power plants, as many predict, become the general, mandatory standard for coal-fired power plants from 2020 or 2025 onwards (so that all future coal-fired power plants are built to this standard), then these power plants will have an operating lifespan reaching to 2070 or 2075 and beyond. Thus they will be subject to the massive price increases expected for coal and gas during this period. This will have a significant financial impact on CCS power plants. As a result of the process of CO_2 emissions capture and transport, more fuel is needed to generate a kilowatt hour of electricity than in standard coal-fired power plants. Installing CCS power plants also involves accepting all the other non-CO_2 emissions produced by coal-fired power plants as well as the dangers involved in the permanent disposal of CO_2, both of which increase over time. The problem of permanent disposal is played down by those who support CCS, just as proponents of nuclear power negate the problem of atomic waste. The term CCS itself is a form of extenuation: "S" stands for storage, and something that is stored is usually being saved for future use. However, CCS involves the irrevocable burial of CO_2, with no intention of it ever being permitted to re-enter the atmosphere.

A key motivating factor behind CCS is that countries such as China, with its coal reserves and rapidly growing electricity needs, cannot be prevented from building new coal-fired power plants. Such needs must be met, and therefore

CCS power plants become added to the catalogue of climate protection measures and given political support – this is seen as climate policy realism. Advocates of CCS claim that the alternative option, of replacing coal-fired power plants with renewable energy, is impossible. However, when one considers the financing necessary for CCS power plants, and the special CO_2 pipeline infrastructure they require, claims that renewable energy is an unacceptable burden on business and national economies are clearly a specious excuse. In reality, CCS technology is nothing more than a life-saving mechanism for large, fossil fuel power plants, and brings with it a multitude of unjustifiable consequences. The CCS option is not a "bridging technology"; it is capitulation to the interests of the fossil energy industry which wishes to maintain the status quo.

Even so, climate research institutes and several environmental organizations believe that CCS should at least be given a try. The WWF claims that, "there's no point damning this technology without first examining it, and hereby potentially squandering the opportunity to protect the environment".[35] But what is there to examine? Certainly it is technically possible to capture CO_2 and send it along CO_2 pipelines to permanent disposal repositories for compression. Trials can only give us more information about the various separation processes and their respective productivity, and about more or less suitable means of final disposal (i.e. *how* CCS can best be affected). But the question is, first and foremost, *whether* we can justify going down this path, and if again there really is "no alternative". There are also chemical and physical reasons to argue against the CCS option, reasons that no permanent disposal solution can change.

Where to store the separated CO_2?

Separated and stored CO_2 is only climate-friendly when it remains in its repositories for eternity. But no one can, or is willing to, guarantee that sooner or later, it won't after all manage to leak into the atmosphere. The arguments surrounding the German government's draft law in 2009 provide indirect evidence of our lack of knowledge, and this is a deficiency that will continue over the short term. Under this law, power companies would only be liable for the secure storage of CO_2 for a period of 30 years, in order to relieve companies of incalculable financial risks. Power companies tried to reduce even this liability period to 20 years. Quite apart from the fact that nothing can compensate for the disaster which would result from CO_2 being set free, these proceedings are horribly reminiscent of experiences with storing radioactive waste.

The latest case in Germany is the scandal over the sloppy storage of radioactive waste in the former salt mine Asse. For decades, research centres and the responsible government institutions referred to expert geological opinion and declared this repository safe. All that was required was that the repository was dry and that there was no chance of water penetrating over the long term.

However, after three decades it was already clear that every day thousands of litres of water flowed into the salt dome. As a result, around €4 billion was needed to retrieve over 100,000 barrels of nuclear waste, to prevent the widespread and uncontrollable radioactive contamination of the groundwater.

After this experience, trust in geological studies which attest that CO_2 can be safely stored indefinitely is incomprehensible. Leaks of CO_2 are seen as "acceptable" when they do not exceed 10 per cent in 1,000 years, which represents 0.1 per mille per annum. No one can guarantee this quota, no one can measure it precisely and no insurance company would accept such risks, with the result that it is again society as a whole which is left to pick up the tab, just as in the case of nuclear power. It would also be naive to assume that CO_2 would escape in only very small and regular quantities. It is far more likely that it starts off slowly, with ever faster and larger doses being released into the atmosphere. This process could be fatal for any creatures breathing air – a "CCS meltdown". A pipeline leak would have similar results. Concentrated CO_2 is an asphyxiating gas; it's heavier than air and displaces oxygen. We have to imagine the quantities of CO_2 we will be dealing with. If we take as our starting point the declared goal of making it mandatory for all-new coal-fired power plants to be CCS power plants, then a 1,000 megawatt coal-fired power plant which captures all the CO_2 it generates would produce 10.75 million tons of CO_2 each year, which is then transported via pipelines and stored. With only 30 coal-fired power plants of this capacity in Germany, and assuming an operational lifespan of 50 years for each, we are talking of 34.7 billion tons – which requires a storage capacity of 34.7km^2.

The Berkeley engineering graduate Ulf Bossel, head of the European Fuel Cell Forum, has compared various methods for the permanent disposal of CO_2 and evaluated them according to chemical and physical laws.[36] One option is to dissolve CO_2 in water. The solution would have a concentration of 3g per litre of water, so that a 1,000-megawatt CCS power plant operating for 50 years would require 254km^2 of groundwater. However, as volumes of underground water flow, they mix with groundwater, rendering the latter unusable as drinking water. CO_2 can only be disposed of by dissolving it in water when water pressure and temperature can be guaranteed to remain constant. But the Earth's many strata make this guarantee impossible, and thus there is the constant danger that CO_2 will be released and make its way into the atmosphere via a circuitous route.

A second option is to compress CO_2 in aquifers (underground strata, usually porous and filled with saline water – so-called saline formations). However, the opportunities to pump water containing CO_2 into these aquifers are very limited as water cannot be compressed. The third option is to fill underground cavities with CO_2, as with compressed air reservoirs or repositories for oil and gas. However, this is not a fair comparison: CO_2 storage is intended as a permanent solution, whereas compressed air, oil and gas stored in this way are continually extracted. Moreover, underground cavities have insufficient volume for storing the quantities of CO_2 we are dealing with and would quickly become exhausted. Added to this, these underground cavities would then no longer be available for the future storage of wind and solar electricity or biogas in the form of compressed air. A fourth option would be to sink CO_2 deep under the seabed, an option which most CCS advocates now reject because here the chance of leaking is greatest and could occur soonest.

Vattenfall, the power company pushing hardest for CCS power plants in Germany, favours compression in saline aquifers and, based on studies by the German Federal Institute for Geosciences and Natural Resources (BGR), sees a storage potential of 20 billion tons, i.e. 20km^2 in Germany. This volume would not be sufficient to "store the entire CO_2 load from at least 60 years of German power plant operations in a climate-neutral manner", as the company claims.[37] But even if it were, the chemical and physical processes outlined above already make the allegedly innocuous nature of this procedure more than a little doubtful.

Obvious cost risks

The CCS option is also questionable from an economic point of view. Currently, modern coal-fired power plants are around 45 per cent efficient in terms of the coal consumed. The laborious process of separating out CO_2 reduces this level of efficiency to 35 per cent. The energy consumed in transporting this CO_2 along pipelines, and its subsequent compression, reduces efficiency further. Therefore, up to 40 per cent more primary energy is required in a CCS power plant simply to produce the same quantity of electricity as in a non-CCS power plant. This increases fuel costs in addition to the extra costs of capturing CO_2, pipeline infrastructure, transport, compression and monitoring. Incalculable rises in the price of electricity are thus inevitable. Even moderate cost estimates today show that electricity produced by coal-driven CCS power plants is no cheaper than electricity produced using wind power. The difference is, however, that the costs of producing electricity using wind power will have fallen by the time CCS power plants have become operational, whether by 2020 or later.

Talking up a project, to get it started and to drum up political support, is a well-known method of making major projects palatable to governments and the public, and in this process advocates of these projects are happy to rely on unproven claims. Proof of the feasibility of the CCS project was given by the Norwegian state-owned enterprise Statoil which separated part of the CO_2 from the gas it extracted, directly at the point of extraction, before immediately compressing it. Thus it was all the more sobering when, in May 2010, Statoil announced that it was closing down the world's largest trial for separating and storing CO_2 at its gas-fired power plant in Mongstad, due to technical problems and lack of cost-effectiveness. Their new electricity and thermal power plant will now commence operations without CCS. Statoil's official reasoning was that "the CCS technique proved far more expensive than previously assumed and in itself would cost more than the entire power plant. It is all much more complicated than we assumed four years ago".[38] At the same time, the Norwegian government (which had been the driving force behind this project, raising €1 billion for development and drumming up support at world climate conferences), announced it would provide no further funding for the CCS project. Even as late as 2007, the Norwegian Prime Minister, Jens Stoltenberg, was equating the importance of the CCS project with that of the US's moon-landing programme in the 1960s.

Cancellation of the Mongstad trial should have been reason enough to drop the CCS approach and, without further ado, to drive forward the transition to renewable energy. Why should we proceed with such a dubious option, one which will be more expensive and requires more time to implement than renewable energy? A CCS option requiring enormous permanent disposal capacities, allegedly available in the quantities required, in itself contradicts the claims that there is insufficient storage capacity for renewable energies, whether in the form of compressed air, biogas, hydrogen or underground pump storage stations.

A phoney analytical compromise

We can hardly expect that the advocates of CCS will draw these conclusions themselves. Too many of them have become involved in the project in the meantime, power companies and governments as well as environmental organizations such as the WWF and even the German Institute for Applied Ecology. The project appeared to be a form of Columbus' Egg, a compromise which accommodated the interests of both power companies as well as climate protection. It is thus a prime example of following the wrong track, despite the object lesson of nuclear power. Almost overnight, the project became a new promise which put politics in motion after, in a "special report" in 2005, the IPCC declared that CCS was an essential element of climate protection policy. Instead of first being scrutinized, the process was immediately integrated into energy predictions and planning. Public funds were provided and legal frameworks introduced, to an extent and at a speed which has never been the case for renewable energy.

In 2008, the EU included CCS as a course of action in its cap and trade directive. With its Climate Change Act, the British government conferred CCS, together with nuclear power, an elevated status as a climate protection measure. This was followed in 2009 by the EU's CCS directive. The German government submitted a draft CCS bill, although it was withdrawn at the last moment in June 2009 amidst fears of a wave of protest immediately prior to the next German parliamentary elections. The Australian government founded the Global CCS Institute with an annual budget of US$ 100 million – a sum more than six times higher than the 2010 budget for the newly founded International Renewable Energy Agency (IRENA) with its 140-plus member states.

The EU provided funding of €2 billion for 12 CCS power plants, Australia US $1.2 billion, Canada US$1.2 billion and the US government US$3.4 billion. Everything was being lined up to recognize CCS power plants as a Clean Development Mechanism (CDM).[39] The largest US environment organization, the Natural Resources Defense Council (NRDC) with its 1.3 million members, became the spearhead for introducing CCS. In doing so, it hoped to drive forward the fight against CO_2 emissions, believing the fight for alternatives to coal-fired power plants to be futile.[40] The Norwegian environmental protection organization, Bellona, saw the closing down of the Mongstad CCS project as a "stab in the back" and moaned that "we've spent a lot of energy defending this policy".[41]

Thus CCS has become a global strategic focus for conventional energy interests. As noted in the introduction, the Shell Corporation has largely suspended its activities for renewable energy, instead announcing its change of focus to CCS. Power companies with their portfolios of coal-fired power plants, and technology companies pinning their hopes on CCS techniques, are jointly financing advertising activities such as the German IZ Klima, an information centre for climate-friendly coal-fired power plants. Another organization, DEBRIV, which represents the German lignite industry, regularly places large advertisements in daily newspapers, in which a series of professors all say the same thing: coal can and must be used in an environmentally-friendly manner with CCS, because renewable energy cannot replace coal-fired power plants. Warnings and protests are explained as irrational prejudices and contrasted with the shared consensus of experts.[42]

Climate research institutes evaluate energy technologies solely (entirely sensibly from their viewpoint) in terms of their ability to reduce CO_2. Even so, in addition to their specialist competency in the matter of climate research, they are also awarded the right to pronounce upon the political and economic strategies which are most beneficial for future energy supply. As the leading climate research institutes are also involved in the IPCC, they automatically adopt the pattern of consensus existing within that community in assessing what is feasible, achievable and acceptable. However, the IPCC is not a committee which bases its recommendations exclusively on the criteria of scientific analysis, although it certainly deserves praise for making the global public aware of the dangers of climate change in the face of organized campaigns of denial. Yet it is reliant on consensus. But this consensus is a phoney analytical compromise, at least when it comes to suggested courses of action. It encourages concessions to be made to influential interests and prevents concrete courses of action from going too far – hence the "two degree" target and the CCS approach. They have published predictions on the economic feasibility of CCS even before reliable empirical data is available. Ottmar Edenhofer, deputy director of the Potsdam Institute for Climate Impact Research (PIK), spoke up for CCS, saying: "CCS can be an important bridging technology for global climate protection. If the technical problems of emissions and geographical storage at competitive costs and acceptable levels of risk can be solved, then according to the PIK's model calculations, by using CCS, the costs of climate protection can be reduced by a quarter."[43] But what is the basis for the calculations in this model, where the technical costs are unclear and the costs associated with potential conflict are ignored?

The potential for social conflict

The CCS project is a dance on top of a volcano. As a survival aid for the coal industry, it is obviously permitted to cost more than renewable energy. The vehemence with which pressure was exerted on members of the German Parliament in order to push through the 2009 German CCS Act speaks volumes. Marco Bülow, in his role as speaker for the Social Democratic Party's parliamentary group on the environment, described in detail how lobbyists continued to intervene in the lawmaking process.[44] Power companies are betting on governments taking over their financial and political risks. Thus the chairman of RWE demanded that the German government assume the costs of constructing and financing CO_2 pipelines, because infrastructure provision is a public undertaking (a point of view he rejects when it comes to electricity networks). Whatever happens, society is left bearing the risks inherent in CCS, as it does with nuclear power. However, it is left to policy to deal with popular resistance to these pipelines and CO_2 permanent disposal solutions. The massive resistance in Schleswig-Holstein, encountered as initial soundings were taken for the first permanent disposal facility, was a foretaste. Similar resistance will crop up wherever CCS begins to be installed. Aware of this, Germany's Environment Minister, Norbert Röttgen, declared that CO_2 permanent disposal facilities would be established only where there was regional acceptance. In a new draft law on CCS, submitted to the German Parliament in July 2010, municipalities housing permanent disposal facilities would receive financial compensation commensurate with the volume of stored CO_2. In addition, this law should be only an experiment and caps annual disposal quantities at 8 million tons.

Talking up the putative advantages of CCS is not only scientifically and economically, but also politically, reckless. Scientists speaking out on behalf of CCS do not have to answer for the investment risks should unpredictable dangers arise. Nor must they win democratic elections in the face of floods of protest from a population which, for good reason, is ever less willing to believe scientific claims of the harmless nature of CCS. The CCS project is based on shifting risk rather than overcoming it, and is an expensive and highly risky means of delaying energy change. Financial reasons and justified resistance ensure that CCS is almost certain to fail. The problem of the CCS project is that it is better at sinking billions in funding than making CO_2 disappear, and again wasting the attention and time needed for the transition to renewable energy.

Recycling rather than permanent disposal

The CCS deception begins with the euphemistic term "storage". Rejecting the CCS approach by no means implies continuing to release large quantities of CO_2 into the atmosphere while we continue to burn fossil fuels, especially as significant quantities of CO_2 are set free during other industrial processes such as the production of cement (CO_2 is set free from the raw material chalk). It is far more sensible to recycle CO_2, i.e. not Carbon Capture and Storage, but Carbon Capture and Recycling (CCR). Thus CO_2 would be transformed from dangerous waste (for which the atmosphere is currently abused as a "wild storage site") into a reusable material. Recycling the products of fossil fuel combustion processes is not CO_2 avoidance, but it does offer us the opportunity to halve CO_2 emissions. One method of recycling is the production of algae. When mixed with CO_2 in a small "algae reactor" similar to a glass container, in just one day and using natural sunlight, algal seeds produce an algal culture which is an industrially usable biomass. The algae yield from one hectare of these reactors is eight times greater than biomass crops.[45] Recycling systems such as these cannot justify new investment in coal-fired power plants, although they could serve to recycle the CO_2 emitted during the production of cement, for example.

Recycling is only practical when the quantities of CO_2 released in a single location are not too large, unlike at large power plants with yearly emissions of 10 million tons of CO_2. As the products of recycling are used over a wide area and in a multitude of different forms, their production needs to be decentralized. This is incompatible both with large power plants and pipeline infrastructures. Again we come up against the key motive in the CCS concept – the determination to hold onto the large-scale structures of energy supply and the structural barriers to energy change and economic restructuring.

The specious baseload excuse

Justification for the continued indispensability of nuclear and/or CCS power plants into the foreseeable future has less to do with the quantities of energy required and much more to do with the necessary "baseload". According to this argument, these power plants are necessary because solar and wind generated electricity is not always available when needed, and to overcome this drawback we would need disproportionately greater, and financially prohibitive, storage for solar and wind generated electricity. This becomes magnified into an apparently insurmountable obstacle for energy change, and so the most inefficient factor in current energy supply – the so-called *baseload power plant* (see p55) – is declared the centre of the electricity universe, and conventional electricity production's last trump card.

The highly lauded benefits of the baseload power plant run counter not only to the structural inefficiency of energy use (due to the unavoidable losses caused by unused steam), but also to the losses involved in transporting electricity and the necessity of further reserve power plants for down times. These are a fairly common occurrence, especially for nuclear power plants. These reserve capacities, required for periods of only a few days or weeks and which represent unused "cold reserves", make up around a third of overall capacity on average. This entirely fails to take into account the reserve capacities available in the form of emergency generators which are rarely used but which need to be available in hospitals, telecommunications facilities, for water supply, rail traffic or administrative facilities. These decentralized power plants have the potential to affect energy change. Hardly anyone has thought of this, even though Germany offers a prominent example of how this could be done: the emergency electricity generator in the Reichstag building in Berlin, the seat of the German Parliament, has been replaced by two combined heat and power generators which run on vegetable oil and cover the building's entire electricity and heating needs. In this case, it is the public grid which functions as the emergency electricity supply source.

No energy supply system can survive without reserves and storage capacities – neither the conventional system nor one based on renewables. In the conventional model, energy is primarily stored prior to its conversion into electricity. If not used immediately, this electricity must be stored. Storage systems include energy transport systems, coal bunkers or tanks. As sun and wind energy cannot be saved prior to being transformed into electricity, storage mechanisms need to be downstream rather than upstream of the generation process. The traditional energy system also stores power, although the need is less as medium load and peak load power plants are usually brought on stream to meet additional demand. Thus the difference between conventional and renewable energy has less to do with the need for storage and much more to do with the form of storage and the investment this requires.

A key element in system change is to replace these baseload power plants with rapidly connectable control energy, by means of intelligent network management and modular electricity production. This reduces the overall demand for storage. The choice of storage mechanism must not depend solely upon isolated calculations of investment costs. The transformation losses associated with storing electricity are countered by the energy losses inherent in a baseload power plant as well as its lack of efficiency and the cost of maintaining the baseload power plant's unnecessary reserve capacities.

To date, renewable energy scenarios have included the storage mechanisms which are used in the conventional energy supply system for meeting periods of peak demand, e.g. pump storage and compressed air power plants. Conventional batteries could be and sometime are also used, yet traditional battery technology's short charging cycles and energy expenditure make this inadvisable. However, discussions on storing electricity neglect the many new battery technologies available, as well as the multitude of new storage technologies which were presented at the annual International Renewable Energy Storage

Conference run by EUROSOLAR and the World Council for Renewable Energy. Most are in the prototype stage or awaiting market launch. Several of these are listed in the section on *System breakers* in Chapter 4 (p105). These storage options sweep away all justifications for building new nuclear or coal-fired power plants, or extending their operational lifespan. Equally, they also wipe out the grounds for installing extensive super grids, as envisaged for the DESERTEC or North Sea projects which I examine in the next chapter. Whether nuclear or coal-fired power plants or the so-called super grid projects outlined in Chapter 3, each is planned with a time horizon of 50 years, each is declared essential for covering baseload supplies and, in each case, electricity storage technologies are spoken of in pessimistic terms. They all ignore the range of options provided by storage technologies for renewable energy – for otherwise the rationale behind these major projects would collapse.

C. MARKET AUTISM: THE FOUR LIES SPREAD BY THE COMPETITION ABOUT RENEWABLE ENERGY

Voices are raised whenever there is any hint of political market interventions on behalf of renewable energy. In contrast, political initiatives in support of conventional energy are seldom derided as being "anti-competitive". Obviously there are two different benchmarks at work here: "*Quod licet jovi, non licet bovi*" – what is legitimate for Jupiter (established power supply) is not legitimate for oxen (renewable energy). In the 25 years in which I have suggested, introduced or accompanied political initiatives to promote renewable energy, the objection I hear most frequently is that these are irreconcilable with free market principles, except when it comes to public funding for research. The recent history of political backing for renewable energy is full of attempts to roll back this support in the name of a "market oriented", or putatively more productive, solution. There are urgent warnings against hasty action and admonishments "not to push too much", because this would "only damage" renewable energy and prevent it from developing in accordance with market forces.

These complaints have been, and still are, brought particularly loudly and persistently by power companies who enjoy a position of market dominance, achieved not through coming out on top in a free market, but as a result of political protectionism. Their blatantly hypocritical demand that renewable energy must be left to establish itself in the market on its own merits is supported by economic research institutes, even those who are not commissioned by power companies to produce studies.

Neo-liberal market correctness

Since the 1970s, there has been no lack of spectacular studies, known to all the decision-makers and pressuring them to undertake comprehensive political initiatives. They include "Limits to Growth" (1972), "Global 2000" (1981), "Our Common Future" (1987), world climate reports and others. However, almost no effective political initiatives have been subsequently taken, partly due to neo-liberal doctrine having become the national and international leitmotif for economic action, and despite growing recognition of the fundamental threats to resource and environmental security. During the 1990s, this doctrine of untrammelled economic liberalization, with its idolization of the "free market" and stigmatization of state-planned economic activity, rose to become the overriding political and economic school of thought. The "invisible hand of the market" was ascribed more rationality than politico-economic strategies.

However, the neo-liberal school of thought unilaterally monopolizes the ideal of freedom for the benefit of business, equating the principle of an "open world" with "open world markets". Since the 1990s, this principle, known as the "Washington consensus", has become the key criterion for judging politico-economic action. Strategies which ran contrary to this principle were listed in an index. Political actions, even when dealing with questions such as resource and environmental security which are fundamental to our survival, were measured in terms of free market principles, with political initiatives for overcoming resource and environmental crises becoming taboo. International economic organizations such as the International Monetary Fund (IMF), World Bank, Organisation for Economic Co-operation and Development (OECD), EU and others followed this doctrine as closely as national governments and political parties did. Even environmental institutes and organizations increasingly strove to come up with projects which did not fly in the face of this doctrine, it being apparent from the outset that anything else would be impossible to implement.

A school of thought becomes dominant when everyone involved regards it as so self-evident that they no longer perceive its contradictions. Just how great these contradictions can be was made particularly clear by the contradiction between two world conferences held just two years apart: at the first, the UN Conference on Environment and Development held in Rio de Janeiro in June 1992 (the Earth Summit), the famed Agenda 21 was adopted, declaring sustainable development and ecological methods of production the central challenges for the 21st century; at the second, the April 1994 conference in Marrakech, the Marrakech Agreement was signed, thereby establishing the World Trade Organization (WTO) and enacting the General Agreement on Tariffs and Trade. This had the effect of declaring global free trade in commodities, capital and services a form of global economic constitution, according to which, henceforth, all political and economic activities should be oriented, enjoying priority even over international agreements on climate or employment protection. Nina Scheer's question "global free trade before environmental protection?" became codified and an "economic century" euphorically proclaimed.[46] Once this order of precedence had been established, it was followed by the demand that political initiatives for the transition to renewable energy adhere to this free market dogma, as if the dogma were more important than energy change itself. Thus the World Bank continued to view the German EEG negatively, even after it had long proved to be the most successful political approach for encouraging the transition to renewable energy. As the same governments voted for both Agenda 21 and for global free trade, this appears to be a form of political schizophrenia.

Claiming adherence to the criteria of market correctness, the dominant players in the power industry are acting as the guardians of the "Holy Grail" of the energy market. After energy markets were legally liberalized, these dominant players belatedly recognized their own unique head-start over any newcomers to the market. Since then they have been playing a duplicitous game: defending their own position, one achieved through political privilege, whilst loudly proclaiming market dogma every time a newcomer appears on the scene. The gigantic quantities of conventional energy with their economies of scale have a huge advantage over the limited volumes of renewables-based power. Yet a decisive economic difference is happily ignored: the more fossil fuels are used, the more their price rises, for these fuels cannot be regenerated and thus supplies become scarce. In contrast, a mass market for renewable energy technologies lowers prices.

A significant feature of neo-liberal economic philosophy is its context-free fixation on the efficiency of isolated products. These are, in turn, compared to other products (again with no regard for context) with the subsequent expectation that they will compete on the open market. Cost comparisons such as these create the impression of ideologically-free, independent precision. Questions regarding the origin, systemic relevance and various social, ecological and economic consequences of each product are suppressed, with the result that this monochrome market philosophy solidifies into an extremely short-sighted ideology, one which stultifies both itself and others and which is ultimately unsustainable. Pressing medium or long term considerations fall through the short-term market net of this ideology. Neo-liberal economic philosophy has the unmistakable touch of autism, and by following its partly rational theoretical conclusions we cannot help but produce irrational results. This has encouraged economists in France to found a "society for post-autistic economics", a long overdue act of intellectual resistance.

We need to resist more actively the myths of the energy market. Permitting like to compete with unlike within a single market runs contrary to the principle of a level playing field. Without a level playing field, free market principles are reduced to a caricature of themselves. Starting from the principle of market equality and the basic requirements of objective comparison, there are four market contradictions between traditional and renewable forms of energy. These contradictions demonstrate that all attempts to promote renewables on the energy market in competition with conventional energy or even, after initial support, to leave it to the mercy of the free-market, effectively function as a brake on renewable energy, and artificially prolong the existence of the traditional power industry. They demonstrate why the criterion of energy cost alone is insufficient to lever traditional energy out of the market, even when renewable energy is, or will be, cheaper.

The myth of competition

Traditional and renewable energy are not competing on a level playing field. For over a century, traditional energy has enjoyed many forms of political support; from direct and indirect subsidies of almost unimaginable quantities, to laws and other privileges. Just look, for example, at Germany, with its mining law and decades of subsidies for coal mines which run into the hundreds of billions; its support for nuclear technology and the Atomic Energy Act, which assumes almost all of the development and liability costs incurred by power companies right up to the present day; the tax-free provisions for permanent disposal, currently to the scale of more than €30 billion, which can be used as the power companies see fit and are therefore effectively tax-free profits; tax-free nuclear fuels and preferential interest rates on construction loans for nuclear power plants; the incomparably generous financial infrastructure of large-scale nuclear research centres; plus members' contributions for Euratom and the IAEA, where German contributions since 1957 alone amount to over €1.3 billion.

Legal privileges range from decades of regional monopolies for electricity and gas supplies which shut out any competition, through to the free construction of networks which smooth the way for operating large-scale power plants and gas supplies. Today political favouritism primarily takes place in the form of tax-free fuel imports. The largest indirect global subsidy is tax exemptions for marine and aviation fuels, representing annual lost tax revenues of more than US$300 billion. All together this has served to concentrate the power industry and permit monopoly structures to develop. This in turn enables monopoly profits which, alone for the four German power companies, run to around €20 billion annually. The liberalization of the electricity and gas markets, introduced by the EU in 1996, has served to speed up rather than to hinder this development. Against this background, demanding that renewable energy should be left to sink or swim on the open market, and damning the laws designed to promote renewables as anti-competitive, is pure cynicism.

Market laws designed to promote renewable energy are a vital precondition for its expansion. They are not an infringement of free-market principles; rather they serve to compensate for the enormous subsidies and privileges which the traditional power industry has received, and continues to receive. They are a means of generating free-market conditions, for they serve to re-establish a diversified range of providers in the market and to satisfy other demands. The only alternative would be to put an immediate stop to the continued privileges of conventional energy and to demand that power companies pay back their subsidies retrospectively. However, neither is it possible to calculate these subsidies with sufficient precision and nor could this retrieval be effected. There are no grounds for a guilty conscience when it comes to market privileges for renewable energy. On the contrary, this is the only way we can compensate for the established imbalances and monopolies.

What we fail to account for: Social costs

Theoretically, competitive markets have a social function because they increase the productivity of providers, reduce prices and encourage the optimal allocation of investment resources. However, this can only apply for products which have a social function. But we have long been aware that prices for nuclear and fossil fuels fail to reflect their "ecological truth" (Ernst Ulrich von Weizsäcker). The established power industry is continually subsidized by society, both today and, because of the long-term effects of environmental damage, over future generations. The magnitude of these subsidies is enormous, although they cannot be precisely calculated and are happily ignored in calculations of energy efficiency. Depending on the costs considered – damages to health, forests, water supplies, land and mountains, a reduction in species diversity, as well as increasing storm, flood and drought damage caused by climate changes – and their taxation levels, they range from ten to 20 times the internal operating costs of producing energy.[47]

Some forms of energy production using renewable energy sources can also have a social cost: this is chiefly the case with biomass, where not compensated for by replanting, and with energy crops grown as monocultures, using unreasonably high levels of fertilizers and pesticides, with high water consumption levels affecting groundwater levels, or the use of genetically engineered seeds. Building large reservoirs, which massively alter landscapes, and redirecting river paths can also create social costs. The key is to use sustainable methods of cultivation to largely avoid these problems, or to ensure that the damage they cause is minimal compared to that caused by nuclear and fossil methods of generating energy. In contrast, the social costs of using traditional sources of energy are significant and unavoidable.

It is paradoxical that renewable energy, with its minimal external effects, is more expensive on the energy markets than our conventional energies with their high social costs. For any other products, discrepancies such as this would have been declared unacceptable long ago. Demanding equal market opportunities for both polluted and clean drinking water, or for contaminated and uncontaminated baby food, would meet with massive protest. Cattle or swine are even slaughtered en masse to protect populations when viral epidemics break out. Such reasoning is foreign to the guardians of the energy market and such measures would be discounted on the grounds that they contradict free market principles. Political market privileges for energy generated with, or without, limited social costs are design elements of a *social market economy*, one which finally takes seriously the *social costs* of nuclear and fossil energies.

Thus the German Renewable Energy Sources Act (EEG), for example, which gives purchasing priority to electricity generated using renewable energy and guaranteed prices which are paid by all electricity customers, is not a form of subsidy as is so often reproachfully claimed. It is actually an *environment bonus* for the marketing of emissions-free and resource-protecting electricity. It is even a means of boosting energy security, because power is produced from local and inexhaustible sources, making it simultaneously an *energy security bonus*. But this bonus will be by no means redundant when, in the foreseeable future, the price for electricity generated from renewables falls to the price of conventional energy. However, demands by many renewable energy advocates for such a bonus are still too limited, for socially damaging energy would then cost no more than socially beneficial energy. The declared economic principle of comparing like with unlike would still remain.

As long as traditional energies are still being used, there must be a permanent, politically supported price advantage for energy which is free or low in pollutants. This can only be achieved using instruments such as the EEG or in the form of tax breaks, equal in size to the social costs which have been avoided. Only then do we arrive at an appropriate social market system. The priority enjoyed by renewable energy in the market must be seen as an integral element of a social market system, rather than an exceptional, temporary solution. Thus all electricity consumers would automatically become eco-electricity consumers, using first the available supplies of electricity generated with renewables before drawing on electricity generated using traditional means. Such a *purchase obligation* is by no means unusual; indeed it is the term *subsidy* which is a deliberate deception.[48]

No one describes energy-saving regulations for buildings, mandatory third-party liability insurance for cars, household insurance, compulsory health insurance or waste disposal charges as subsidies, whether for the construction industry, insurance companies or waste disposal institutions. Each is invariably regarded as a public good or necessity from which we all benefit equally, and thus it cannot be left to the individual to decide whether to contribute financially or not. This is particularly the case for renewable energy. Market theoreticians who fail to recognize this are living on a different planet. What we call neo-liberal is actually an unqualified attempt to solve macroeconomic problems using a microeconomic approach. In the power industry, this type of approach generates high social costs which are not included in the price of energy, and severely damage societies' living standards.

Protected oligopoly

No one can seriously claim that the nuclear and fossil power industry functions according to market principles. Before there were any discussions about liberalizing energy supply, their regional monopolies gave them the status of risk-free, planned economies. The power companies primarily saw liberalization as an opportunity to extend their business activities geographically. According to the European Commission, with its goal of a single EU-wide electricity and gas market, there was anyway only room for seven providers in the EU's internal market. In order for these providers to compete within the European market, consolidation would be needed within the electricity market (but although the separation of electricity production from transmission networks and distribution networks was stipulated, there was no requirement to demerge ownership structures).

Only a few member states, such as Sweden and the Netherlands, have taken the step of putting networks into public ownership, preventing them being abused by electricity producers who use transmission costs to create barriers for their competitors. Other self-serving privileges can be achieved by refusing, or delaying, extensions to networks in order to disable competing providers who need network access to supply their customers. When the liberalization laws were drawn up during the 1980s and 1990s, no one imagined that many new providers would appear on the scene in unusual geographical locations. At the time it was still inconceivable that renewable energy could be a significant competitor to conventional energy supply.

At least an electricity market has gradually developed, one in which electricity consumers can swap between providers. However, this development has been matched by the simultaneously accelerating process of concentration. State monopolies such as the French Electricité de France (EDF) have been able to expand into neighbouring countries. In Germany, large power companies have bought up regional and municipal suppliers. Thus, despite the legal dismantling of regional monopolies, the dominant players have even been able to strengthen their market position. Before network charges were controlled by a single regulatory body, electricity providers attempted to maximize their competitive advantage by lowering their own electricity prices and simultaneously raising network charges. After this option was curtailed in Germany by the Federal Network Agency (BNetzA), which was founded in 2005 (later than in other EU countries), electricity prices rocketed. In order to maximize profits, new investments into power plants and maintaining networks were reduced. The result of this development is ageing power plants and a neglected network, a pattern seen with all privatizations, even those outside the field of energy supply such as the railways and water supplies.

This development couldn't be a permanent one. It unavoidably lead to a conflict in which the power industry is now faced with the decision of either to adapt to the demands of energy change, or to hold onto its existing structure by making new investments into new large-scale power plants and networks. It is the municipal power companies above all who are seizing this opportunity to become active participants in structural change, for they see in renewable energy and the expansion of heat and power cogeneration the opportunity to take up the role of producer once again, a role from which they have been ejected over the past decades. In contrast, the major power companies are betting on the perpetuation of a system of large-scale power plants and associated networks, in addition to producing electricity from renewable sources in a form which suits their own structures, e.g. large-scale projects including offshore wind parks or large-scale solar power plants.

The established power industry is being duplicitous. On the one hand it speaks of energy markets, when the market mechanisms can be used to hold down alternatives, to buy up competitors and to expand. On the other, it cries that governments should be prevented from supporting potential competitors – "for the benefit of the whole", although really only for their own benefit. This is an audacious demand when seen from the perspective of a market economy, and one which many completely fail to perceive having mentally accepted the dominance of the power companies.

The "planning security" demanded by electricity producers is nothing other than their desire for political protection of the status quo. Thus their behaviour patterns can be defined as follows – as much planning as possible, only as much competition as politically necessary, and at a level which is beneficial to the electricity producers themselves. Any real competition is excluded because of the role of power companies as global players; this is a role encouraged by politics and secured via international mergers and acquisitions. Commercial exploitation should, as far as possible, extend right from energy extraction through to the transport of fuels, to power plants, refineries and the subsequent transport of useful energy. Here it is the oil companies who have the upper hand, companies which themselves are largely owned by the oil-producing countries. The three fundamental errors of the "market-based" liberalization of energy supply are clearly recognizable: the first was the failure to demerge the power industry, which would have been the compulsory prerequisite for creating a functioning energy market. The result was a process of concentration on ever fewer providers, at a speed and extent previously unknown. Second, everyone ignored the fact that conventional energy reserves, which are geographically limited, finite in quantity, and tied up with international infrastructures and long-term, large-scale investments, were too inflexible for free market conditions. And third, the fact was overlooked that no government in the world can risk a breakdown in energy supply, for then everything would come to a standstill. A "free energy market", based on conventional, non-renewable energy and energy imports, is an illusion.

The cost trap of traditional energy supply

Evidence for this includes the price increases for oil, gas and coal over the past few years, increases which were higher than the slowly increasing costs of extraction. As the individual costs of extraction necessarily vary, the prices set by primary energy providers reflect the relatively high international extraction costs. Prices are set at the highest rates that consumers can pay without risking a collapse of the market. The traditional power industry is perforce international because reserves are finite, and thus it finds itself with a "natural monopoly", one which can only be overcome through the transition to renewable energy. The oil extraction countries and supply companies are able to accumulate ever more capital which is in turn lost to the recipient countries. In turn, the extraction countries and supply companies, largely state enterprises, buy goods from companies based in recipient countries – far outside the realms of the power industry. Thus there are shifts in the balance of the world economy. Energy-efficiency analysts who fail to recognize this are professionally blinkered. When has there ever been a provider with an eye on profit margins and a market monopoly who has been satisfied with price levels that merely cover costs and offer limited yields?

The transnational power industry, which can only exist as a protected market economy, grows not in spite of the approaching exhaustion of conventional energy reserves, but because of it. During the dawning era of the conventional energy system, the size and influence of the primary energy industry grows, shaping its entire future. They who own the source have incontestable power over the market, a power which can only be broken by military intervention in oil extraction countries (as, for example, during the Iraq war), and where participation in the market is only possible through close cooperation with the relevant government or through corruption. Oil extraction countries who work together, in their own interests, and reject the option of rapidly exploiting their resources, can keep prices high by keeping supply low and simultaneously saving their reserves – as does the Organization of the Petroleum Exporting Countries (OPEC). Demands from oil-importing countries that they increase supplies and lower prices betray an embarrassing helplessness and lack of political imagination.

We can only escape from this historical market trap by making the transition to renewable energies which do not need to be imported from other national economies. And only by switching to renewables can the provision of energy truly be organized according to market principles, because the sources are inexhaustible and the technologies can be duplicated. However, for the reasons set out above, this transition cannot take place within the current energy market, with its fundamental and only partially revisable inequalities between conventional and renewable energy.

D. THE LACK OF POLITICAL MORAL COURAGE: PLAYING OUT THE FUTURE TODAY

The alternative to the centralized monostructure of current energy supply is a multitude of largely decentralized structures. It is no surprise to anyone that established power companies try to use every measure possible to hold onto the current system. It accords with their premise of pushing for a slowdown in energy change and invoking actual constraints, although in truth these are efforts to maintain the system. Their arguments, only superficially plausible, appeal to habit, mental inertia, a lack of information, indifference and undefined fears of anything new. In taking this stance, they are attempting to play-off the future against the present.

Therefore they plead volubly for international consensus, and warn against rapid unilateral action which purportedly leads to international isolation. But why should it not be possible, using quickly installed renewable energy systems, to match France's achievements with nuclear power plants that require long construction times? From 1977, the year in which the first nuclear reactor went into operation, to 1987 (i.e. merely a decade), France brought 39 nuclear power plants into service, which generated 50 per cent of its entire energy supply at that time. Whatever one thinks of nuclear power, France did not propel itself into a position of international isolation by doing this. Warnings that renewable energy "also has its environmental problems", expressed in order to gloss over the substantial dangers associated with nuclear and fossil energy, are equally duplicitous. By claiming that everyone is a sinner, enormous problems are equated with minor ones, as if, metaphorically speaking, there were no moral difference between a capital offence and a case of pick-pocketing. Suggestions that there is "also resistance to renewable energy" among the population runs along the same lines, as if the reasons for such resistance were as significant as those against nuclear or fossil energy systems and as if this resistance was equally widespread and well-founded. According to opinion polls, in Germany at least, less than 10 per cent of the population favours the construction of new coal-fired or nuclear power plants, and more than 90 per cent support the rapid expansion of renewable energy. Over 60 per cent would accept a wind power plant within view of their homes.[49] Even so, reports are happily made about local resistance to wind or water power plants, while at the same time expecting governments to force through the construction of large-scale power plants and high-voltage networks in the face of all resistance – in "the general interest".

Particularly perfidious are the notorious warnings about higher prices for renewable energy which in turn will endanger "supply security". These warnings are designed to drown out all discussions about the urgent and predictable dangers associated with nuclear and fossil fuels, and to prevent them being recognized. Multi-billion dollar power companies, some with extremely high yields, behave like lawyers for social concerns. These attempts to encourage egotism at the cost of other people and future generations is an insult to society, implying that the majority are prepared to risk catastrophes for fear of change.

Yet despite the continual stream of publicity, these methods are becoming less and less effective. The more often renewable energy is seen in action, the more popular it becomes. In a survey asking whether "having energy produced in the neighbourhood is very good or good", 74 per cent replied "yes" where energy is generated in solar parks, 56 per cent for wind parks, 40 per cent for biomass systems, but only 6 per cent for coal-fired power plants and 5 per cent for nuclear power plants. Almost 5,000 people were questioned in this survey.[50] In Germany, everyone knows that mobilizing renewable energy by means of the EEG has led to an increase in electricity prices, and will continue to do so in the next few years. Compared to the overall price of the customers' electricity, these increases are relatively low although they are regularly denounced as excessive. Even so, the EEG has become ever more popular because renewable energy represents our hopes for the future and because people understand that temporary, additional costs serve to drive energy change forward. Thus negative propaganda largely backfires.

With this in mind, we have to ask what causes and misleads governments, parliaments and political parties into paying more attention to the needs and interests of the established energy system than their actual vulnerability permits. Despite avowed commitment to renewable energy and even with willing engagement on its behalf, established centralized structures continue to be seen as pre-eminent and indispensable, as if the decentralized structures of renewable energy were simply a childish phase that we will outgrow. Attitudes such as these can be found even among active advocates of renewable energy in energy and environmental science, and in environmental and renewable energy organizations.

There is no single, common denominator to explain this. The reasons are also psychological. For some it's ignorance, others underestimate the technological advances. For some, vested interests obstruct the view for coherences, and others are unable to think in terms of structures. For many it is perhaps an unconscious, fundamental trust in established energy providers, and internalized respect for the magisterial heritage of conventional energy supply – even though this is the main cause of our ecological global energy crisis. My impression is that there are three overriding motives for accepting the premises of the established energy system.

The first was described by the political scientist Martin Greiffenhagen in his study of the psychological advantage enjoyed by incumbents, upon which political conservatism relies. Anyone calling for system change is expected to prove something for which there is no concrete proof – to show that something new that does not yet exist is better than something which is apparently tried-and-tested. This describes the monumental psychological advantage enjoyed by established energy corporations, one which engenders the knee-jerk reaction of forgetting that anything which has the appearance of being tried-and-tested has, by definition, no future. The established energy system with its large-scale structures is deceptively appealing to our need for security, and encourages a fear of alternatives. These structurally conservative tendencies are also inherent in the scientific system. Scientists, wishing to avoid speculations to protect themselves from accusations of lack of scientific basis, often prefer to focus on existing structures rather than accepting new framework conditions and examining new technologies. Indirect evidence of this is seen in their many published studies and scenarios.

The second motive is the assumption that energy change can only take place hand-in-hand with the existing power industry, and thus must fit in with existing structures. It is "unrealistic" not to do so. This is a viewpoint particularly widespread among political stakeholders. "If you take on the power industry, you won't get anywhere," is something I have been frequently told – usually by politicians who have not done so themselves and thus have been able to change almost nothing. However, it is an empirical fact that there have been huge successes in introducing renewable energy against the interests of an established, and apparently indomitable, energy system. That was political realism. You can only accelerate at full speed when you're not tied down and having to consider the constraints of your current situation (i.e. old investments not yet amortized and other system investments). We cannot even calculate a theoretical point in time at which all traditional energy investments will be simultaneously written off, for these investments are staggered and of varying lifespans. The traditional power industry's systemic capacities are a mix of older and more recent investments and thus can only follow the strategy of gradual transition, a strategy which always results in putting off that which could actually be realized now. Hence it is only the protagonists for change who are not integrated into the current system who can serve as the driving force for energy change. It will be their initiatives that force energy corporations to adapt to new developments. Initiatives have only ever been successful when they run contrary to the system, and when new activists in politics, business and society take control.

The third motive results from the suggestive power inherent in the words "large" and "small". We only take large seriously – a *large* project, a *large* power plant – rather than the many small initiatives, even if their joint effect is greater. Therefore it is essential to recognize what is "large" and "small" in terms of energy change. Large demand is not synonymous with large power plants. Although it is correct that only large projects can drive forward energy change, it is misguided to equate this with the construction of large power plants. An example of a large-scale project made up of numerous independent individual initiatives is the village electrification programme run by the West Bengal Renewable Energy Development Agency (WBREDA) in India, which provided 3,000 villages with electricity generated using renewables within five years. A second example is that of the Grameen Shakti Bank ("Shakti" means "the sun") in Bangladesh which started up in 2004 and, with its micro-credit and maintenance services, will have introduced 1.5 million photovoltaic systems, 100,000 biogas plants and 5 million solar cookers by 2012. A third example is the German EEG which encouraged investments worth €96 billion into generating electricity using renewables during the period 2001–2010. This is in contrast to new investments for conventional large-scale power plants of less than €10 billion during the same period.

The continued adherence to conventional large-scale power plants, and even new plant construction, is consistently justified by the claim that these are required to cover energy demand over and above the limited role played by renewable energy. This assumes the validity of various scenarios which regard the potential share played by renewable energy as limited. In German policy discussions on energy, this is the reason given by the conservative Christian Democratic Union/Christian Social Union (CDU/CSU) party and liberal Free Democratic Party (FDP) for reversing the 2001 decision to phase out nuclear energy, and authorizing the construction of new coal-fired power plants. Their left-wing counterparts, the Social Democratic Party (SPD), continue to favour the phasing out of nuclear energy, but justify their acceptance of new coal-fired power plants with the argument that it's not possible to simultaneously give up both. And the Green party too, as the next chapter shows, believes we cannot do without large-scale structures although these should be used with renewable energy and super grids. Yet the logically consistent answer to the question of how to cover our entire energy needs without reverting to nuclear power, coal-fired power plants or CCS is obvious: a policy designed to massively accelerate expansion of renewable energy. This would include the necessary storage options for which the natural potential and technologies are already available and to which, by consistently implementing the principle of priority for renewable energy, supply structures must be adapted. Such an adaptation must also enable the system functions to be decentralized in a multitude of ways. This decision differs from the "overall energy policy" which is so often demanded, one which attempts to stipulate the share to be played by renewable and conventional energies in the form of energy industry planning for decades ahead. Any plan that attempts to do this will fail because it cannot help but diverge from actual economic and technological developments in energy.

It is not only the interests of the established power industry, embedded in political institutions and parties and often in the form of individual persons, which prevent political activists from introducing the structural revolution in energy supply. Such methods of entrenchment have been known about for a long time. Particularly effective has been a form of legal corruption along the principle of "we'll pay later", in which government members and leading civil servants, after leaving their posts, are subsequently offered well-paid jobs with energy corporations. This explains a lot, although by no means everything, as the circle of people thus affected is relatively small.

Of greater relevance is the lack of political moral courage, the attempt to avoid inevitable conflict and to continually search for "energy consensus". But one can only bring together that which fits together. However, society has moved much further on than most of its political representatives realize. Everyone in a political role must ask themselves whether they believe that paralysing agreements with established energy interests are more important than the objective necessity of driving forward energy change, and the legitimate expectations of society that we do so, with all its implications and without delay.

3

SUPER-GRIDS AS PSEUDO-PROGRESSIVE BRAKES

DESERTEC and the North Sea Project, the new megalomania

Two gigantic plans for renewable energy, the DESERTEC Project and the North Sea Project, have captured the imagination of the public, simply as the result of media hype. These two projects are designed to supply Europe with electricity generated by solar power plants in deserts and by offshore wind parks in the North Sea, with the electricity being transported to Europe via a transcontinental network of new high-voltage transmission lines. This network is praised as a "super-grid", essential for delivering power which will be produced entirely from renewable energy. Taken together, the projects appear to provide a panacea for overcoming all the seemingly insoluble problems of solar and wind-generated power.

DESERTEC/Transgreen

The DESERTEC project, presented in July 2009, aims to build solar thermal power plants and wind parks in the Sahara and the Near East to generate electricity which is then transmitted to European countries over distances of 3,000–5,000km along specially constructed power lines. The DESERTEC project aims to meet 15 per cent of Europe's power needs by 2050, with 3 per cent already being covered by 2020. Project costs are estimated at around €400 billion, making it the largest investment project the world has ever seen. A key argument for the project is that it would enable us to overcome the problem of storing solar electricity; the planned Concentrated Solar Power (CSP) plants can generate electricity continuously, and will thus function as so-called base-load power stations, as defined in Chapter 2. This is technically possible because the sun shines constantly in desert regions, and because the basic principle behind CSP plants is the same as for all conventional power plants – they produce steam which drives turbines to generate electricity. During the day, the heat they generate can be stored in saline tanks and used to produce electricity at night. The technical design is impressive. As these storage systems enable electricity to be produced at all times, CSP plants are *the* key element in the DESERTEC project. The project's initiators predict a price of only €0.05–0.06 per kilowatt hour for the electricity it generates, making solar electricity unrivalled in terms of cost. Building the transmission networks and power plants would also generate jobs in European industry. At the same time, the project aims to drive forward the process of energy change and to function as a form of economic development aid for countries in this area that will profit from the electricity exports.

The project has generated wide support and, in addition to Munich Re, has gathered under its wing organizations and companies who have long been considered antipode, and still are when it comes to decentralized concepts for renewable energy and adherence to nuclear and fossil fuels. They include power companies such as RWE and E.ON, the Club of Rome and Greenpeace, technology corporations such as Siemens, who continues to build nuclear and fossil fuel power stations in addition to its new business of renewable energy technologies, as well as several companies that specialize in solar technologies. The political and media response has been almost universally positive. The EU Commission has guaranteed its support, as has the German government and, irrespective of their other conflicts regarding energy policy, most German political parties (the CDU/CSU and the FDP, the Greens and also, with slightly more reluctance, the SPD). Many were under the impression that, at last, here was a project on which everyone could agree, and allowing a line to be drawn under all previous controversies. "Today we have promised to save the world," was the statement made as the consortium was officially presented.

The DESERTEC project has a strong "German focus", both in terms of the consortium's membership and the industrial and power industry interests it articulates. Thus it was inevitable that DESERTEC would challenge other initiatives being undertaken by EU member states who regard the Southern and Eastern Mediterranean as their own particular sphere of interest, both for reasons of geography and history. This is especially the case for France for the Francophone part of North Africa, its former colonies. Thus under the aegis of EDF, the French government responded during the spring of 2010 by presenting its own "Transgreen" project. The key to both projects is a high-voltage direct current (HVDC) electric power transmission network. HVDC networks involve transmission losses of only 3 per cent over a distance of 1,000km, compared to 10 per cent in an alternating current network, and are thus known as "super-grids".

North Sea Project/Seatec

Super-grids are also the basis for the second project, the North Seas Countries Offshore Grid Initiative, or Seatec. Presented in January 2010, the media resonance was comparable to that of DESERTEC and the project generated broad support. Involving not only the EU Commission but also nine governments of EU member states, Seatec is riding the wave of euphoria generated by DESERTEC. As well as connecting all the North Sea countries to pump storage power plants in Norway, the primary aim of the project is to continue to develop offshore wind parks in the North Sea so that they form the future focus of European wind power generation. The European Wind Energy Association (EWEA) and even Greenpeace are supporters of this plan. Norwegian pump storage plants will be used for bridging windless periods in Central Europe and Great Britain, with the power produced by offshore parks during periods of intense wind being used to pump water back into the storage plants. As the offshore wind parks can rely on permanent wind availability, the network can guarantee supplies of wind-generated power over a wide area and bridge periods of regional calm. The costs for this North Sea super-grid, stretching over thousands of kilometres, are estimated at around €30 billion.

Ten European corporations have come together to form a group they call Friends of the Supergrid. They include several manufacturers of wind power plants, the French nuclear technologies corporation AREVA, Siemens, and the RWE-subsidiary HOCHTIEF. Together they are demanding that the network be financed by the EU Commission and national governments. They plan to install an offshore wind power park with a capacity of 100,000 megawatts linking to the super-grid, which is 50 times the current installed offshore capacity of Great Britain (688 megawatts), Denmark (663 megawatts), the Netherlands (247 megawatts), Sweden (164 megawatts) and Germany (72 megawatts).

Premature euphoria

Both projects, but especially DESERTEC, have been enthusiastically acclaimed, and generate the impression that only the backward or narrow-minded could seriously have objections. The prospect is being held out of linking both projects, as well as joining up with Transgreen. The positive reaction to both projects is not due solely to the major success this represents in promoting renewable energy; they also promise to overcome the systemic conflict between conventional and renewable energy, and the structural conflict between central and decentralized systems of power production, as well as feeding hopes of a possible energy consensus, one which may permit the interests of power corporations to be reconciled with the ambitions of the environmental movement. Cooperation, integration, historical compromise – each is coming together. This is the dream of politics and business, and we feel we are in safe hands.

However, all the deeper motives which underlie the reservations about renewable energy, as described in the last chapter, remain relevant here: the tendency to structural conservatism, the fundamental trust in large-scale power plants which has developed over the past decades, and an underestimation of the potential offered by a decentralized system of power supply. The scale of this underestimation becomes clear when we look at the differences in perception between these large-scale projects and the major successes booked by the EEG, which get overlooked in our premature euphoria over desert and offshore power. Both projects appear to be the long-awaited "grand design", and they push the greater triumphs which have already been achieved from public consciousness. If a quarter of the electricity produced by DESERTEC were to flow to Germany to cover 15 per cent of Germany's power needs then, according to the model of cost estimates, Germany must invest €100 billion into the project. If we compare this with the €96 billion spent since 2000 on investments into generating electricity with renewable energy, made in response to the EEG and enabling us to cover 13 per cent of our power needs within ten years, then it is clear that the "major project" of the EEG has been significantly quicker to realize and is by no means more expensive than the DESERTEC project promises to be. Incidentally, this €96 billion represents innumerable small and medium-sized investments, and is impressive evidence both of the speed at which the public wishes to see renewable energy introduced, and of how investment capital for renewable energy can be raised.

The excessive importance given to large-scale projects was also evident in an article by Fritz Vorholz, editor of the highly-respected German weekly broadsheet *Die Zeit*. Writing of the North Sea Project, and describing the first large-scale German offshore wind park with its 70 megawatts of installed capacity 100km off the German coast, he declared that this marked the beginning of "the electrical future" – as if Germany's current 25,000 megawatts of "onshore" installed capacity hadn't already long since marked its beginning.[51] Another, even more significant, difference of perception is that between the DESERTEC project and the "100 Per Cent Regions" conference being held in Kassel at the same time. At the conference, 99 German administrative districts and cities presented their own plans (many already being implemented) for covering 100 per cent of their power supply needs by means of renewable energy by 2015, 2020 or 2025.[52] After these plans have been realized (and many are already far advanced), these districts and cities will have no need for solar electricity transported across continents – even before any such solar electricity becomes available. Whereas the newspapers were filled with talk of DESERTEC, media response to this conference was practically zero.

Therefore it came as a surprise to many when I described the DESERTEC project as a *fata morgana*, and the North Sea Project as a distraction from a faster way of making the transition to renewable energy. I was asked how anyone who is regarded as an advocate of renewable energy, and who pushes for complete energy change, could be against a project like this. However, I am by no means alone in my criticism. Yet in the thunder of applause accompanying the super-grid projects, welcomed and lauded throughout the whole energy sector, the only voices submerged beneath the din were those raising serious objections. This group included the bulk of activists who have accomplished the feat of setting us off on the path towards renewable energy. They see little reason to accept that conventional energy corporations, who to date have been primarily concerned with blocking strategies for renewable energy, should suddenly be the driving force behind its introduction, irrespective of their interest in maintaining the status quo. What is the real motivation for these projects, and where do attempts to make them a key aspect of European electricity generation lead us?

A. SUPER-GRIDS: PROLONGED DETOURS ON THE PATH TO RENEWABLE ENERGY

The super-grid idea is well-received by all those who cannot imagine that energy change can be realized in a different, more effective, faster, more economically productive and socially beneficial way, despite exemplary object lessons to the contrary. It also goes down well with everyone who hopes to avoid systemic conflict by arguing that in future we need both centralized and decentralized structures, without admitting that this creates inevitable conflicts of interest, targets and priorities.

An argument against these large-scale projects is that, where it is made dependent upon these projects being realized, energy change is unavoidably slowed down. Where energy change takes place independently of these projects, then the gigantic super-grid threatens to become a financial black hole. The same applies to the super-grid approach which is, if anything, even more centralized than the traditional energy system. The assumption is that the super-grid will create a structure for delivering renewable energy, although renewable energy is, by its very nature, decentralized and thus already available at the point of consumption. By proceeding with these projects, we are choosing to take a major detour although direct paths to energy change are available. The projects are purely designed to meet the needs of the power industry, rather than the economy as a whole, and pay absolutely no attention to the needs of regional economies. They reduce, rather than increase, the number of players involved in renewable energy.

These super-grid projects adopt the premises of the traditional energy system and turn these idiosyncrasies into a fixation on the – supposedly permanent – necessity of geographically concentrated power generation. Although described as revolutionary, the concept behind the super-grid project bows down to the premises of an obsolete model. Accepting these premises involves declaring that, although not yet built, the super-grid is the prerequisite for a change in the system of power supply. Any delays in its construction, or indeed its failure to be built, are thus a justification for energy corporations to continue to operate their current conventional large-scale power plants. However, if the super-grid were to become reality, then power companies, with their huge organizational and capital strength, would be able to build power plants both in deserts and offshore and, in doing so, to transfer their dominant role as producers into the renewable energy sector. For the power companies, the super-grid approach is an *à la carte* design. It allows them to continue with their conventional methods of producing electricity at least until the super-grid is built - and this could be a very long time.

Yet bear in mind that we are not dealing here with the question of whether electricity produced by renewable energy should be directed through transmission networks or not. This can make sense and be necessary in order to balance power supply between regions, with excess power production in one region being used to compensate for a temporary lack of production in another. This requires some extensions to existing networks and more connections between them. However, the super-grid involves the construction of a completely new high – and highest possible – voltage network, intended to run from the North African and Arabic desert regions right through into Central and Northern Europe. The Irishman Eddie O'Connor is regarded as the creative force behind a "direct current highway" for Europe,[53] and Günter Rabensteiner, chairman of the management board at APT, Austria's largest power trading company, and board member of the European Energy Exchange, talks of a "European copper plate", with the aim of "trading power between nations on the basis of a unified European market model for a harmonized system of power trading".[54] That would certainly be advantageous for the power trade, although not necessarily the right approach for affecting energy change. A project of mammoth proportions such as this would have direct practical consequences for the entire transition to renewable energy. It would largely be placed in the hands of the established power corporations, the only party with an interest in drawing out the transition to renewable energy for as long as possible. The planned power plants can only be built and commissioned when the super-grid is actually available. Yet the time required to construct a new network of the intended dimensions is significantly greater than that needed to build the power plants themselves. Thus the effect would be an inevitable delay of incalculable duration.

Furthermore, a super-grid for desert and offshore power plants is synonymous with location planning. It limits the numbers of locations and leads to the spatial concentration of production – and thus to a commercial concentration of capital and the spatial concentration of value creation. It accommodates projects that promote the growth of renewable energy only in areas with above average levels of solar radiation or wind, making European wind power primarily a feature of the Atlantic coastline or offshore, with solar power primarily in southern Europe, in North Africa or in the Near East, as recommended by the Potsdam Institute for Climate Impact Research (PIK). This ignores all the opportunities for linking natural sources of renewable energy directly to regional energy demand, an arrangement for which there is far greater economic and social rationale than simply the cheapest possible method of generating power. Worse still, it may even indiscriminately hinder these opportunities, in favour of the theoretically "greater" rationality of energy management.

B. TECHNOLOGY WITHOUT SOCIOLOGY: THE INCALCULABLE DESERTEC PROJECT

When it comes to the DESERTEC project, only the name and the details are new. The idea of supplying Europe with energy generated from solar power in the Sahara is 60 years old and was developed by the Frenchman Marcel Perrot at his solar research institute in Algeria: electricity generated by solar power plants in Algeria, then a French colony, was to be transmitted along cables lying on the seabed, thus securing future French power supplies. After France was forced to recognize Algerian independence following the Algerian War of 1954–1962, this idea was rejected and France started out on its nuclear journey. The same idea had already been seized upon during the 1950s by the German professor Eduard Justi who wanted to generate hydrogen, using solar power plants in the Sahara, for supplying Europe with power. This idea was picked up again in the 1980s by the German Aerospace Center (today's DLR) and developed further. This research institute has been experimenting with solar thermal power plant technologies in Almería in Spain since the 1970s, and developed the Transmediterranean Renewable Energy Cooperation proposal upon which the DESERTEC initiative is based.

Doubtlessly the DESERTEC project is technically possible. Yet compared to other renewable energy options, and despite all the optimistic studies, its true cost is still written in the stars. However, because of its fundamental importance, we cannot allow the actual cost of energy change to be the key criterion, at least not when renewables are compared with nuclear and fossil energy. The question of cost is a relevant one though when comparing various means of utilizing renewable energy. In addition, costs should never be considered in isolation, because supplies of renewable energy must also involve other, and more important, economic and social benefits, especially the question of which renewable energy concepts can be realized fastest, both for Europe and in developing and transitional economies.

Unrealistic schedule, exploding costs

For political, economic and sociological reasons, it is clearly almost impossible to realize the desert power project whilst adhering to the promised budget and planned schedule. In saying this, I do not wish to imply that errors have been made by those who drew up the relevant studies, but studies such as these can, as best, only calculate what would be possible if everyone involved in the project followed the plans to the letter for a period of several decades. This is unrealistic, the number of institutions involved alone is unmanageable; more than 40 governments, each with their own national network systems, and not forgetting the many regional governments across whose land the power lines must be laid (and therefore also the populations of these regions who cannot be permitted to disrupt construction).

No major technical projects have been successfully completed within budget – costs are often twice or even more over budget estimates. Even when correctly calculated, the complexity of major projects makes it impossible to accurately predict time spans or the actions involved. No study can anticipate the outcome of decision-making processes, levels of resistance, conflicts of interest which arise and unavoidable errors occurring as the project is implemented. Major projects can lead to major errors which end in a debacle. The desert power project would be the world's largest investment project to date, with an unpredictably large and (due to the length of the project) continually changing number of participants. It is overly complex and ignores the delicate political situation in the majority of the affected desert states. Having a realistic view of the political and economic environment is far more important than computer simulations of technical costs.

Moreover, it is hardly possible to draw up cost estimates for the long-distance power lines with any degree of accuracy as to date they are unique. However, EUROSOLAR has had similar projects scientifically evaluated: the longest HVDC lines in the world are in China and stretch over a distance of 2,000km. In contrast, DESERTEC is planning power lines of lengths up to 5,000km. To date HVDC lines have been point-to-point connections, thus a densely branched network with many cross linkages – as is planned for DESERTEC – will involve greater development costs. This desert project represents a jump in at the deep end; we have no experience of using HVDC cables at depths of more than 1,000m, as will be needed to cross the Mediterranean. This alone makes it extremely unlikely that the first two HVDC lines will be in place by 2020, quite apart from the fact that it is daring to calculate such a gigantic project before the key element upon which it is based has gone through its own learning curve.

But above all, it is extremely unlikely that the construction of transmission routes will go according to plan because local resistance is inevitable. For the long transmission routes through Italy, over the Alps to Germany, through Spain over the Pyrenees or through the Mediterranean to France, and from there to the various connection points, the likelihood of there being individual pockets of resistance is incalculably large. The problems encountered in building many transmission routes of only 20km or 50km in length, often taking decades to complete, is a foretaste of what is to come. An example of the type of problems DESERTEC will face with cross-border network construction is the conflict surrounding a French-Spanish high-voltage line running from Cazirel-Aragon, to Baixas, Santa-Llogaia. Although covering a distance of only 75km, construction has gone on for 30 years. Only after former EU Commissioner Mario Monti had been engaged as a mediator and with the promise of more than €225 million in EU funding could agreement be reached, allowing construction to be completed by the year 2014 and in the form of expensive underground cabling, just to placate local objections. The total costs for this power line are around €800 million, nine times the initial estimate!

One of the main arguments raised by civil protesters regions through which the DESERTEC power lines will run will be that the lines are not designed to meet their regional, or often even national, power needs. It is conceivable that conflicts over transmission routes – and we are talking of 80–100 routes, each of lengths between 3,000km and 5,000 km! – will be a permanent feature of the project. Even where all conflicts are solved to the benefit of the project, significant delays in construction are unavoidable and this automatically creates dramatic increases in costs. This cannot help but have consequences for the subsequent construction of power plants which rely on the availability of the transmission routes.

The DESERTEC project is fixated on CCS power plants for they are vital if the fundamental principle behind the project is to be realized – to make desert electricity attractive as a means of covering European demand for power as long as the structures are designed to cover baseload demand. If this project were primarily concerned with generating solar and wind power for countries in North Africa and the Near East, then it would also need to consider, in addition to CCS power plants, photovoltaics and wind energy. These are not reliant on a trans-Mediterranean super-grid and could be introduced relatively easily over a wide area. CCS power plants are of value to major cities in the desert regions of North Africa, or as balance reserves for wind power in the particularly windy regions along the Atlantic coastline and in large parts of Egypt. In contrast, photovoltaics are of great value to the many rural regions across North Africa, where network connections are almost impossible as well as unnecessary.

In addition, only air-cooled, rather than water-cooled, CCS power plants can be used in desert countries because of the lack of available water and/or the immense additional costs of desalination plants and water pipelines. Air cooling is used in the world's largest CCS power plant currently under construction – the 100-megawatt project Shams 1 in Abu Dhabi. However, this project shows just how far CCS power plants are away from being able to produce electricity for €0.05 or €0.06 per kilowatt hour – including transmission costs to Europe. In order for the Shams 1 power plants to be amortized, compensation of €0.30 per kilowatt hour over 25 years has been agreed with Abu Dhabi's network company. The amortization costs for Abu Dhabi's next 100-megawatt plant, currently in planning, have been reduced to €0.25 per kilowatt hour. These two power plants, being constructed to meet the local demands of Abu Dhabi consumers, are not subject to any land rents and involve only minimal transmission costs. For purposes of comparison, guaranteed kilowatt hour compensation for these two power plants is higher than payments for power generated by photovoltaic systems in the open countryside in Germany.

Investment conflicts

As the desert power project can only be realized in cooperation with the governments of the countries in which it is located, these governments must also have a financial stake in the project. Where the national power companies in these countries are unable to build power plants themselves, the spectrum of possibilities for generating returns ranges from rental income (which is then reflected in production costs) through to an equity stake in the power plants, and export taxes. Another possible option would be for the host countries to receive a share of the electricity generated by the foreign power plant operator for free, or at a reduced price. It is inconceivable that the countries hosting the power plants and transmission networks will simply provide the locations without expecting to participate in the returns they generate. Indeed, we have to assume the opposite – the more the recipient countries depend upon these power supplies, the greater the share expected by the countries hosting the power plants. Just look at the oil-exporting countries in many regions which were formerly colonies. Step-by-step, and after achieving independence, these countries fought for an increasingly large share of the returns, until finally nationalizing the extraction companies and then dictating prices.

A far more important, and obvious, factor in the economic development of these desert countries than a share of the returns is the rapid reduction of their fossil fuel imports. These represent an extremely high economic burden, one which can be alleviated through the transition to renewable energy. This is certainly the case for the majority of the desert countries with no national oil and gas reserves. But it requires that the production of solar and wind power be entirely focused on the needs of these countries, and not those of Europe. This is the only way to divert the focus on the potential of desert power into a more constructive direction, both for the desert countries and for Europe. But this strategy demands an entirely different approach, one I outline in Chapter 5 and call the "Desert Economy" – a terminological allusion to the DESERTEC project yet one which diverges in substance. I regard my approach as an initiative for fundamentally refocusing DESERTEC towards the needs of the countries in which the project is based, because the many problems described above make it inevitable that the project in its current form will be impossible to implement. All that is currently being presented as DESERTEC's initial success is plans for power plants in North Africa to cover local demand – projects which had anyway already been conceived and planned and which, for years, EU funding programmes have been available but not fully utilized.

Technically feasible is not the same as socially feasible

Just a few weeks after the DESERTEC foundation had been established and the consortium publicly presented, the initiators were forced to admit that it would not be possible to achieve the proclaimed target of supplying 3 per cent of European power needs by 2020: the investment funds had not been secured, and nor was a route plan for the power lines ready for authorization. The delay was explained by the need for further studies and the search for North African partners. To date, the project consists primarily of very successful publicity which is aimed at motivating governments to contribute financially, especially for constructing the transmission network. It is likely that by 2020, no single HVDC line of the intended length will be in place. As a network consists of more than just a few power lines, the first power supplies could be routed to only a limited number of recipient countries, maybe only a single country, which would then be expected to be particularly committed to the project. The DESERTEC project may well have been started, but it will never be completed. Some companies will wish to take part, only to back out sooner or later once confronted with the actual circumstances.

The project is technocratic and completely ignores any sociological factors. Technocratic designs typically state that everything that is technically possible can be realized, while dismissing questions and references to potentially adverse economic and social circumstances as speculative. Their promoters are happy to point out that implementation is merely a matter of "political will". Thanks to this get-out clause, the responsibility for misjudgements and failures lies with others and are never the fault of the project's designers themselves. It is unrealistic to expect that the planned timescale of four decades for achieving the project's targets can be adhered to – it is highly likely that more time is required, with a gigantic plan mutating into gigantic strategic blunder. When justifying their structurally conservative behaviour, the creators of the DESERTEC project simply ignore the many methods of storing renewable energy, a topic I examine in more detail later. The same is true of power companies – for then the only remaining justification for their refusal to meet all power demand with renewable energy, and reasons for transmitting electricity over great distances via a super-grid, would no longer be valid. This is also the argument behind the Seatec project.

When it comes to the desert power project, the primary motivation for the power companies involved is not the low electricity prices, which the studies estimate at €0.05–0.06 per kilowatt hour for Europe. These attractive numbers are simply a welcome justification for denouncing continued growth in decentralized renewable energies as both uneconomic in principle and irresponsible – long before the first long-distance power line from North Africa to Europe has been built and the first solar power plant can be connected. In contrast to the beliefs of the study's authors and an impressionable public, it is unlikely that the power companies truly believe that the DESERTEC/Transgreen project can be realized in its entirety, and let alone at these low costs. One can hardly accuse the power companies of a lack of practical experience; rather they are attempting to use these promises to win the systemic conflict in their own countries between established power supply systems and renewables-based power supply. Above all, they win time by doing so. However, France may have additional motives for taking part: to use the North African-European super-grid for delivering electricity produced in nuclear power plants which AREVA hopes to build in North Africa. The European MEP Claude Turmes speaks of a "hidden nuclear agenda" in France's official involvement in the trans-Mediterranean desert power project.[55]

C. BREEZY CALCULATIONS: THE ECONOMIC CONSEQUENCES OF SEATEC

Proof that the primary concern of power companies lies not in low-cost renewable energy but rather in maintaining their own structural dominance can be found in their preference for building offshore wind power plants. Here the costs are permitted to be higher! This preference for offshore wind power cannot be justified on a cost basis, for offshore wind power will always be more expensive than its land-based counterpart. The first German offshore projects were only undertaken after the feed-in tariffs specified in the EEG were twice significantly increased. They are currently around €0.06 per kilowatt hour higher than for onshore wind power plants, and an even higher guaranteed price is being demanded. In Denmark, the guaranteed price for onshore plants is 0.5 krone per kilowatt hour and this for a total of only 22,000 hours operation; for offshore power plants the price lies between 0.52 to 0.98 krone and for 50,000 operating hours, i.e. in total around three times higher! Even so, the reasons given for expanding offshore production were that wind power could be generated more cheaply because of more favourable wind conditions at sea. However, due to the significantly higher installation costs, at no time was this claim tenable. Installation costs include the building of plant foundations at depths of 40–60m below sea level, as well as greater maintenance and network connection costs. Yet what may be economically senseless when compared to onshore wind power projects makes perfect sense to the established power companies: offshore parks can only be established as a joint effort, with several wind plants being installed all at once, because undersea cables are prohibitively expensive if laid only to serve individual power plants. Thus offshore wind parks are a popular playing field for large investors.

Yet when municipal utilities, such as those in Munich, participate in such offshore projects, this is by no means a contradiction. Munich's municipal utilities aim to supply all of Munich's private households with power generated using renewable energy by 2015. Without wind power playing a major role, this is not possible. However, the more obvious method for producing wind power more cheaply – wind parks located in Munich's direct environs – remains politically impossible due to the Bavarian State government's negative policy of refusing to authorize the use of almost any sites for this purpose.

Evidence of the wrongheaded ideas which the Seatec project encourages is demonstrated in the plan to use offshore wind power for pumping water into Norwegian reservoirs – rather than using power generated by wind power plants sited along Norway's long, and largely empty, coastline with its excellent wind conditions. It is no doubt sensible to use Norwegian hydroelectric power as reserve or balance energy, to compensate for fluctuating wind conditions in European countries, and to build the necessary cable connections to Central Europe in order to do so. This requires only network connections between existing hydroelectric plants and connection points within existing networks. However, this does not justify shifting the focus of wind power production out to sea.

Another, frequently suggested, reason for favouring offshore wind parks is local resistance to onshore wind power plants. Yet surveys continually confirm that this much touted resistance does not represent the opinions of the majority of local populations. The book *Wind des Wandels* (*Winds of Change*), which I edited together with Franz Alt, documents this clearly on the basis of pertinent examples and opinion polls. Power companies who justify their preference for offshore parks by referring to this type of resistance are untrustworthy: when it comes to the construction of new coal-fired power stations, which are rejected by more than 90 per cent of the German population, then they expect political institutions to push these projects through in the face of all resistance, just as they do for the construction of new high-voltage power lines. The "price" for renouncing or rejecting the expansion of wind power in the regions, and of refusing to install onshore wind parks, is new power lines for gigantic projects: if you refuse land-based wind power plants you get super-grid power lines!

D. A CONFLICT OF PRIORITIES: THE POLITICAL MISUSE OF SUPER-GRID CONCEPTSTO FIGHT DECENTRALIZED POWER PRODUCTION

As we have seen, super-grid plans are a symbol of the continued structural conflict between centralized, network-dependent power production and its decentralized alternative. At the political level, this creates a conflict of priorities over the path to renewable energy. For a super-grid project representing new investments to the tune of tens of billions, the *minimum political precondition* is public funding contributions from the EU and/or the transit countries. Where governments meet these preliminary conditions, we can expect that the next step will be a limit on investments in renewable energy which does not need super-grid connections, so as to aid refinancing and capacity utilization of the power lines. Every government has the necessary instrument for doing this – the refusal to authorize planning permission for renewable energy sites. This would thwart or kill off decentralized investments made by multitudes of small investors (the fastest method of realizing energy change), either in countries in which desert power is produced or in the recipient countries, possibly in both. We can expect the same reaction when the super-grid is privately financed; large investors will make financial contributions dependent upon government assurances of network capacity utilization.

An example which should serve as a warning is the Tennessee Valley Project, carried out as part of President Roosevelt's "New Deal". It involved the construction of large hydroelectric and coal-fired power plants which could only operate at full capacity when connected to long overland power lines stretching into the previously unconnected, rural Midwest. Almost all the farms in this area already had their own wind power generators; several million of these "small wind plants" were in operation. After the government dictated that everyone should be connected to the network, these small plants were shut down.

Thus if the decision is taken to go ahead with DESERTEC/Transgreen, and public funding is made available, we can expect increasing pressure on governments and parliaments to limit the expansion of solar and wind power generation in their own countries and to phase out or abandon laws stipulating feed-in tariffs. Although there is no mention of limiting the decentralized expansion of renewable energy in the plans presented for the DESERTEC project – its stated aims being for desert power to make a limited contribution to power supply and one congruent with decentralized power generation – the originators of this project cannot prevent established power companies using it as a political weapon to fight the expansion of local and regional solar and wind power generation overall. They won't be stopped by recommendations that "only" 15 per cent of the nation's power be imported from the desert.

Already DESERTEC's initiators are bemoaning the fact that feed-in tariffs for renewable energy in EU member states are only valid within the national boundaries, i.e. only for "locally" generated power. Although current transmission capacities are minimal, there are already demands that these feed-in tariffs be expanded to include solar or wind generated power imported from desert countries. This is a direct attempt to replace laws on feed-in tariffs in EU member states with import quotas, quotas which could also be met using production plants in other locations – similar to the system of carbon trading and not linked to actual power supply. This would be a huge step backwards for the transition to renewable energy, making it subordinate to conventional power companies' plans for building power plants.

To argue that a super-grid concept supports rather than hinders the rapid expansion of decentralized power supply is pure nonsense: it forces renewable energy into the structures of established power supply. It has the character of a planned economy for European power, one to which solar and wind power generation will be forced to adapt for decades hence. Once renewable energy has become an integral component of the energy market – a point which will soon be reached, thanks to the successful expansion of wind power generation in northern Germany – then further expansion of the super-grid must wait. This would lead to an incalculably long delay in the super-grid's expansion, one which those currently supporting the thesis that the super-grid is absolutely essential would find hard to contradict. As the super-grid approach is based on the assumption that decentralized power supplies generated using renewable energy cannot satisfy baseload demands, then the opposite is also true: as soon as continuous power supplies can be secured with renewable sources of energy, then the super-grid is surplus to requirements. The super-grid concept represents the incipient conflict over the structures of renewable power supply. It will also influence the conflict over the focus of network investments – whether they should be used to expand the super-grid or to strengthen existing networks for locally and regionally generated power.

Seen from their own standpoint, it is understandable that established power companies favour a super-grid concept. In contrast, from the viewpoint of all those who regard energy change as extremely urgent, it is incomprehensible that this concept be adopted. A super-grid will unavoidably constrict the route to renewable energy and disregard its technological potential, as well as significantly reducing the number of those involved in its development, i.e. it subjects a popular movement to technocratic control. I do not claim that this is the intention of all those who have rushed to praise the DESERTEC or Seatec projects. However, we need to think through the consequences of these projects: the "energy world" does not consist only of supporters of a rapid transition to renewable energy, and within its circle of support are widely differing ambitions and economic interests.

Part II

People, scope for creativity, and technologies for 100 per cent renewable energy

4

SPEEDING UP

The free development of renewable energy instead of technocratic planning

"Energy revolution" has become a fashionable term. Certainly, energy change can only occur by means of a technological revolution. History shows that such revolutions are sparked off by a new and fundamental technology, one which satisfies needs previously unmet or not yet recognized. The new technology then stimulates the growth of new applications and boosts production levels. This in turn leads to an increase in the technology's productivity, stimulating the creative design of related products and applications which themselves encourage further leaps in innovation. However, a technological revolution is not the consequence of the technology itself but of the consumers who grasp the new opportunities being offered. What starts as a technical novelty becomes a social movement, extending into all areas of society, changing its norms and practices, and setting new cultural standards. Every political, economic, technological and social transformation has occurred in this way. Such developments can be stimulated "from above", although this is rare; they can only develop "from below", once accepted by society.

A transformational process such as this cannot function with industrial-scale technologies. It only works with technologies used by a large proportion of the population, giving them greater freedom to act and thus generating mass demand. The most recent example of this process at work is information technologies which, in just a short space of time and with a scope and speed previously undreamt of, have unleashed economic structural change and a socio-cultural revolution which, directly or indirectly, permeates all sectors of society and affects the way we live and work. This has been the consequence of three wireless technologies: satellites, laptops and mobile phones, whose use continues to grow. Initially failing to understand that information technologies offered consumers the promise of autonomy, the telecommunications monopolies initially tried to prevent their spread and to keep their customers linked to cables.

The "satellite" we call the sun also sends its "energy information" all over the world, free of charge. But to date there have been far too few systems available to receive this information. Again, only a few fundamental technologies are needed to push forward energy change and usher in its attendant social transformation. Such technologies make possible that which currently appears impossible, in the light of today's tangle of mutual and complex practical constraints. The precondition for any rapid change is "reducing complexity", as the sociologist Niklas Luhmann describes it: the widespread establishing of new facts, without first having to ask permission of the established institutions. The mass introduction of these technologies restructures the entire power supply system, compelling a reordering of the status quo.

Against the ushers

Any revolution descends into farce when the time, methods and actions of the "revolutionary forces" are dictated, have to be announced in advance and advertised. The farce becomes all the greater when authorization procedures are determined by those who are anxious to stop the revolution in its tracks. And the term "revolution" is finally reduced to a paradox when used by conventional power companies who try publicly to suggest that they are organizing a revolution against their own interests. When authorizations for site locations and investments are also required to adhere to international regulations and are at the discretion of central authorities, then the ironic statement that "the revolutionaries are requested not to walk on the grass" springs to mind.

The technological revolution in energy supply, in which renewables will eventually supplant conventional energy, can only be realized through innumerable, independent and widespread initiatives, and will not succeed as the result of technocratic planning by political and economic decision-making elites who organize according to time schedules and spatial hierarchies. This is especially the case in our digital age, in which the speed at which products replace one another has increased. In his book on the changing time structures, the sociologist Hartmut Rosa describes the "fundamental difficulty" in which the simultaneous "perception of elevated goals of change on the one side" is counteracted by "subcutaneous torpor on the other". This torpor is "particularly evident, above all in verbal communication and the immunization of systemic operational logic". This results in "motionless motion", a "rushing standstill", in which "nothing can stay as it is without some key element having to change".[56]

There is hardly a more fitting example of this diagnosis than the one we are dealing with here: on the one hand we have the accelerating development of modular renewable energy technologies, and on the other the established power industry, the most structurally conservative sector of the modern economy (an inbuilt structural necessity), with investments in exploration, infrastructure and power plants stretching half way into the new century. However, when it comes to energy change, the key question will be who determines how these revolutionary technologies are used, their room they have to manoeuvre and the aims of their activists. The smaller the number of activists and the more that current interests have to be accommodated, the slower the speed of energy change. And equally, the greater and more varied the number of activists, the more quickly and extensively this change can be realized.

Technologies enable us to use the sun as a gigantic source of energy to satisfy all mankind's energy needs, either directly or after its natural transformation into wind, water, waves or plant life. Thus we are dealing with technologies of existential importance for civilization's future. All the problems and difficulties, conflicts, arguments of scale, timeframes and ways in which energy change can be organized, are the result of two fundamental characteristics of solar energy, and four specific ways in which these technologies can be used.

The instrumental characteristics of renewable energy

Renewable energy has two economic characteristics: it comes free of charge and is available wherever it is needed. To this is added its inexhaustibility and pollutant-free nature, something to which no one could object. However, it is precisely the cost-free and ubiquitous natural availability of renewable energy which makes it so threatening to the traditional power industry. Energy debates which fail to recognize this are sham debates.

Nature has already determined that renewable energies will win through in the end. The primary energy industry, its existence based entirely on fossil reserves and uranium, will disappear from the scene altogether, either earlier than it is willing to accept, or too late. The outcome of the conflict surrounding the second key attribute of solar energy, its ubiquitous nature, has yet to be decided: whether mankind's energy needs should be met by local sources of natural energy, or if renewable energy will be appropriated in specified, concentrated areas and then delivered over long distances to consumers, along the lines of conventional power supply structures. As described above, opinions are divided over the question of structure. Although rarely mentioned, it is this dichotomy which has shaped the controversy surrounding renewable energy right from the beginning, as demonstrated by the numerous administrative hurdles to establishing production plants for renewable energy. Now openly discussed, this question has shifted to the centre of our attention. Concentrating production systems in specific locations maintains centralized structures of supply. In contrast, the widespread production of power using renewable energy represents a broader, more comprehensive definition of an energy industry – with a national and regional economic focus, and communal or even individual autonomy – i.e. the opposite of that intended by the conventional power industry.

For those concerned only about a change of energy source, environmental reasons in general and secure, long-term energy supply, these structural questions are of little interest. But for conventional power companies, fervently trying to hold on to their habitual role as centrally organized producers and suppliers, they are of great interest. They are also of interest to those who support regional production and value creation, those with organizational interests or a direct, personal interest in emissions-free energy supply. Industry too, which benefits from a rapidly growing market for renewable energy technologies, should stand up for its spatially unrestricted development and thus for a system of decentralized power supply.

Renewable energy technologies have four specific qualities which determine that decentralized production is not only the faster, but also more economically-efficient and socially attractive option:

- *Energy is appropriated and transformed using a single technical system*: the appropriation of solar radiation and wind power and their immediate transformation into electricity in *one and the same* plant is a unique technological simplification, one which opens up a multitude of autonomous applications;
- *Energy appropriation, storage and use in a single geographical location*, which enables an energy economy to develop based on short distances and the step-by-step renunciation of large-scale infrastructure;
- The use of both *small- and large-scale* solar power systems, where smaller systems are no less efficient because increases in technological productivity occur during their production and not during their use;
- Energy technologies can be *integrated* into products that we already need, thereby incurring only minimal additional costs: e.g. the glass front of a building which is simultaneously a solar module.

These hybrid opportunities apply to many renewable energy technologies and some are already being used today. They cannot be quantified using standard calculations of energy-efficiency. Taken together with their specific qualities, renewable energy's two economic characteristics indicate that the catalysts for energy change are the potentially innumerable autonomous users who benefit economically at municipal, regional and national level, and the manufacturers of systems for utilizing renewable energy.

Anything that breaks up the rigidity of the current system and can be realized immediately serves to accelerate this change. The pace of change accelerates when the technologies involved are not over-complex, are easy to install and use, and are modular. A large number of independent investors are standing by, ready to invest. This is impressively confirmed by the figures given at the end of Chapter 2, which are the result of many small individual investments rather than a few major technical projects. If Germany had decided upon a project à la DESERTEC in the year 2000, rather than its EEG, then it is highly likely that, by the year 2010, no single additional kilowatt hour of power generated by renewable energy would have flowed into the German supply network.

A. SYSTEM BREAKERS: THE GROWING TECHNOLOGICAL POTENTIAL FOR AUTONOMOUS ENERGY

Renewable energy technologies are instruments of acceleration. The technological potential for using renewable energies autonomously is growing constantly, with its potential applications becoming ever more varied. This development, and the increased attention it is receiving, stimulates further, application-oriented developments which in turn increase the intelligence of the technologies and their potential applications. This results in what Helmut Tributsch refers to as "solar bionics" – the functional intelligence and efficiency of natural systems becomes the model for technical developments.[57]

Think of the spectrum of new possibilities for generating electricity being opened up by photovoltaics, some of which are already going into production: organic-based solar cells with micro- and nano-structures, which save on material, increase cell performance, and can be easily and flexibly installed in the form of plastic cells or in pigments. Or solar cells with concentrators, soon ready for the market, which already double efficiency levels and thereby significantly reduce costs. Or windows that produce solar power. In future, every horizontal or vertical construction surface, every roof and the front of every building can be used to generate power.

Think of the spectrum of possibilities offered by solar thermal methods of power generation: the development of thermoelectric solar cells which drive combined heat and power plants; small-scale solar thermal power plants; the decentralized use of saline storage systems, storing the sun's heat captured using parabolic reflectors; magnesium hydride systems in which the sun's heat is used to release hydrogen from the magnesium compound, with the H_2 being temporarily stored in containers before being fed into a Stirling engine to generate electricity; the use of highly concentrated heat for industrial manufacturing processes; every greenhouse can be heated with solar heat or by heat pumps powered by solar electricity; the potential for cooling systems to use solar heat recovery techniques; organic plastics for solar collectors, heat reservoirs and heat exchangers; long-term heat storage systems using silica gel as a storage medium.

Think of the large and hugely productive wind power plants dotted around the landscape, as well as the largely ignored potential offered by small-scale wind power plants, metropolitan wind power generation in urban canyons and between skyscrapers – a concept already in operation in the Gulf state of Bahrain. Or the hugely underestimated potential of small hydroelectric plants in a multitude of running waters, including systems that can be anchored like boats. Or wave power plants utilizing the energy of ocean currents, ocean thermal energy conversion plants, and floating solar platforms in coastal areas.

Think of the energy potential offered by biomass from agricultural residue and organic waste in cities; the innumerable energy crops – not food crops – including water plants, even salt-water plants, and especially algae; the possibilities offered by the dual use of open land for both agriculture and solar and wind power generation; hydration processes using organic materials and enzymes to produce bioethanol; the production of kerosene and petrol from organic waste.

Think of thermal potential, not only deep geothermics but also ubiquitous surface geothermics, as well as heat in the air and bodies of water – bearing in mind that Stirling engines can be driven by low temperature heat; of the many engine types which can generate power using a whole range of fuels – biogas, vegetable oils, bioethanol and even synthetic fuels; of the many types of equipment which can cover their own power needs by means of integrated solar cells and micro-motors.

And think of the many opportunities for constructing zero-emission houses using these technologies; organic forms of insulation; houses which can source all their necessary energy – from heating to cooling and even power for electric automobiles – solely from the natural energy in their surroundings, which alone could cover up to 50 per cent of society's energy needs. And finally, think of the storage potential and control techniques touched on in the previous chapter, for use in demand-driven network management in area, island, local and regional networks; of municipal and regional "smart-grids" which drive "virtual power plants" by managing the energy flow between widely scattered plants, a system already successfully practised by several public services and generating considerable energy savings. The mass introduction of electricity meters together with time-dependent tariffs, giving electricity consumers the power to control their costs and optimize their patterns of use – public services in Sacramento, California are currently making this possible for all of their 600,000 electricity customers. And let's note what Amory B. Lovins and his team at the Rocky Mountains Institute in Colorado, after painstaking calculations, refer to as "small is profitable": that we must finally find a new way of making calculations about energy, one that considers not only avoided fuel and external costs, but also the avoided costs of transporting energy.[58]

Now compare all this with the huge dimensions of large-scale projects, involving highly complex, widespread renewable energy networks and infrastructure with its associated dependencies, opaque processes, technological vulnerability, unmanageable number of anonymous players and profiteers – i.e. technocratic planning strategies that demand the international and streamlined fine-tuning of all energy investments. This is mistaken energy planning, a plan designed to meet a single criterion – that of the apparently optimal allocation of energy investments, greatest possible cost efficiency and lowest possible prices. However, plans such as this only ever work in computer simulations and not in real life where members of society have differing motives, priorities, values and interests.

Allowed to develop freely, renewable energy technologies will inevitably become the determining force because of the manifold, autonomous ways in which they can be applied. Current and future technologies for harvesting, transforming and using renewable energy, from the smallest to the largest and with differing degrees of autonomy, are simultaneously the catalyst for more social wealth distribution, production and economic structures. In comparison, conventional large-scale power plants are inefficient and inflexible outdated models; even new large-scale power plants are a form of technological underdevelopment.

New storage and reserve potential

Some of these technological options enable energy generation to be combined with energy storage. A plethora of other options also exist. For decades little was done to improve battery performance due to apparent lack of demand. This only changed with the growing need for micro-power storage systems for mobile phones, laptops and hybrid cars. Powerful, wireless communications equipment would never have been so successful without the development of extremely light and long-lasting high energy density electrochemical storage mechanisms. The imminent mass production of electric cars guarantees further leaps in development. The lithium ion battery currently receiving such attention is not the only new electrochemical storage technology; others include the redox flow battery and the sodium-sulphur battery. Some batteries already have 100 megawatt capacity.

From a practical standpoint, it is absurd to claim that comprehensive storage capacities must be installed before it is possible or responsible to expand the use of renewables: investments in power storage capacities will only reflect actual levels of power generation and are designed only to meet existing storage requirements – and not vice versa. When there is concrete demand for storage capacities, investment will follow. This is the process behind every technological development: absolute need encourages investment and stimulates further innovation.

As contributions to the annual International Renewable Energy Storage Conference (IRES) show (a conference run by EUROSOLAR and the World Council for Renewable Energy),[59] new storage technologies are ready for market launch: large batteries that enable entire areas to be supplied with continuous power, generated purely from solar or wind power; thermal reservoirs for high and low temperature heat; underground storage; flywheels with longer revolution times; the targeted use of cogeneration for electricity needs. One of the simplest methods of storing power is already being practised in Denmark: in the form of hot water which is then used in Stirling engines to produce electricity and requires temperatures of only 70°C.

The creativity behind developments in energy storage technologies is demonstrated by a technology presented at the IRES in 2009 for the first time: synthetic methane, produced from hydrogen and carbon dioxide. In contrast to pure hydrogen, methane can be easily fed into existing gas networks for distribution. Methane can be transformed into power and heat by means of cogeneration, and also used to power automobiles. Equally interesting is that, in addition to its broad spectrum of uses, methane is also suitable for short-term network stabilization as well as for seasonal storage, something that must be guaranteed where supplies are based exclusively on renewable energy. Micro-cogeneration plants can provide reserve energy, and this is the target of the 100,000 micro-cogeneration plant programme established by Volkswagen and the German green electricity company LichtBlick. The same applies to hybrid systems which combine to use wind power, biogas and solar power – synergy power plants driven by renewable energy.

Solar power generated on buildings and freestanding roofs, and wind power produced alongside motorways and rail tracks, can be used to charge the batteries of both electric cars and locomotives. The growth in electric cars simultaneously brings a huge storage capacity, effectively free of charge, with the storage system already a component cost of the automobile. Thus electric cars can be productive even when stationary. This dual use option will set a precedent and spark a cultural change similar to that generated by laptops: the solar power system on the roof, financed as a roofing cost, and power storage, financed via the car. Who wants to stop the market explosion this will lead to, and by what political means? Are investments in new large-scale power plants still of any value?

Renewable energy technologies are "system breakers". The smaller the dimensions and larger the storage capacities, the less infrastructure we need, with all its associated costs and dependencies. Small format applications are more attractive to the many people for whom energy economy is by no means the only motive, and these in turn stimulate a flood of investments. Analyses of energy efficiency reduce everything to the question of whether each form of energy "pays off", as if we all adopt the same calculations and values. Clearly this does not reflect reality. Look at cars: if all drivers were concerned only with utility value, i.e. driving performance versus purchase price and maintenance costs, then by now there would only be one model of car on the market. The entire automotive industry could save the billions spent on advertising their products and limit its activities to publishing technical data and product price lists. Both the advertising industry and the designers would be out of a job. In spite of the fundamental differences between conventional and renewable energies, it is thus in no way plausible to argue that, when choosing sources of energy, decisions are made solely on the basis of current costs. This argument demonstrates the conceptual restrictions within which many energy economists operate.

From the passive to the active energy society

In contrast, what we all want is access to secure supplies of energy. In developing countries there are billions of people whose only concern is securing access to any form of energy, especially electricity. When this access is made possible using renewable energy, even via methods different from those traditionally available, then increasing numbers of people will grab this opportunity – and the cheaper it is, the faster they will do so. Where renewable energy can be appropriated by autonomous means, it mutates from a pure economic and consumer commodity into a cultural commodity. That is the social logic of renewables: they stimulate the shift from a "passive energy society", with ever fewer and increasingly large-scale providers serving a unified market of planned energy consumers, to an "active energy society", in which energy supply is increasingly autonomous and appears in a multitude of new guises.

This shift is supported by other economic factors. As the primary energy is free of charge and permanently available, the cost of generating power using renewable energy is easier to calculate than with conventional energies, where fuel costs constantly rise. This mathematical advantage also extends to the modular availability of the technologies: where use of these technologies is decentralized, then local capacity needs can be calculated precisely, avoiding the error of investing into overcapacity. Where demand for capacity grows, additional modules can be easily and quickly installed. And as installation times for decentralized systems are very short and production begins at once, equity redemption also begins immediately.

All this makes it clear why the mushrooming potential offered by renewable technologies thwarts long-term energy plans, whether for conventional or renewable energy. Long-term investments become unavoidably risky, a risk that can only be limited by continuing to deliberately block renewable technologies by refusing to grant authorizations, prohibiting use, making connection and use compulsory, direct and indirect subsidies to providers of centralized structures, right through to import restrictions for new energy technologies in countries in which they are not produced. The risks facing large investors can be reduced only by means of political intervention. If renewable technologies are allowed to develop freely, investments in large-scale plants and widespread infrastructures could become a disaster over the medium-term, because capacity utilization levels will continue to fall.

The fact that power plants will become uneconomic to run and need to be decommissioned is actually a relatively minor problem, for only the power plant operators themselves will be affected. A more general problem is that of widespread infrastructures, from gas and oil pipelines through to high and highest voltage distribution networks for power transmission, where regional production reduces demand for long-distance supply routes. Customers who remain dependent upon the under-utilized networks will inevitably be forced to bear a greater share of the transmission costs. This in turn strengthens the trend to autonomous power supply, making maintaining the network even more expensive. As renewable energy technologies continue to rapidly develop, the conventional energy supply system will reach a tipping point, triggered by changes at the end of its supply chain, even before it becomes obsolete as a result of conventional energy resources becoming exhausted.

To prepare ourselves for this development, we need what the sociologist Oskar Negt calls "sociological fantasy and learning by example": an increasing number of "best practice" examples develop into a widespread social movement.

B. PLAYERS: THE SOCIAL AND ECONOMIC MOVEMENT TOWARDS RENEWABLE ENERGY

A movement is created by initiatives which have a broad impact. Although the initiatives must be organized, once a movement has been established then no further input is required. This is certainly the case for movements which result from an idea that is hugely attractive to wide sectors of the population, and whose social value is as indisputable as is the necessity of putting it into practice. Once there is clear proof that an idea can be realized, it becomes increasingly difficult for its opponents to let it play itself out. They can only put obstacles in its way, using increasingly questionable methods.

The key driving force behind energy change is the greater social legitimacy enjoyed by renewable energy. The power of this legitimacy is indirectly indicated by the verbal commitments and greenwashing methods used by those who still wish to halt its advance, whether by permitting or specifying a restricted role for renewables within the traditional energy system. The German government justifies its "comprehensive energy policy" by announcing that the "share played by each energy supplier" needs to be newly determined. To concede to this, instead of playing the trump card of renewable energy's legitimacy – its greater social value – is the biggest mistake that renewable energy advocates could make.

It is also a mistake to concede to context-free cost comparisons with conventional energy, or to concur with the claim that political strategies are only concerned with making renewable energy more competitive. It is not wrong to continue to lower the price of renewable energy – but it is wrong to focus entirely on costs. It is a mistake to agree to energy policies that set quotas for renewable energy, or to emissions permits for fossil power companies which effectively establish market shares in the form of "pollution rights", because this relativizes renewable energy's greater legitimacy. All the "time and target" concepts, demanded and drawn up at world climate conferences and set out in national power strategies which set quotas and deadlines for the introduction of renewable energy and the continued use of conventional energy, are examples of such mistakes. Although these concepts are intended as a means of making energy change mandatory, they simultaneously serve to legitimize the continued use of conventional energy and thus encourage its public acceptance. However, it is nonsensical to fix quotas for renewable energy because the social movement towards renewable energy has taken on a life of its own. Its manifestations are many and various. It doesn't speak with one voice and so can't be a negotiating partner. A forced deceleration in the take up of renewables would be comparable to the political regulation of renewable energy. Many who speak up in favour of quotas for renewable energy fail to consider this consequence.

Emancipation from the conventional energy system

The social movement towards renewables develops in a series of practical steps, emancipating it from the conventional energy system with its perforce technocratic patterns of decision-making and hierarchical decision-making procedures. This social movement is the result of numerous bottom-up, rather than top-down, initiatives. Emancipation is never bestowed, nor is it something to be decided upon; instead it must be practised and experienced. It instigates, produces and cultivates freer forms of behaviour. The aim of energy emancipation is to liberalize renewable energy initiatives. It is not the result of government intervention "from above", with society simply looking on. Every government is enmeshed in a network of interests and this is one of the reasons for the political inertia towards renewable energy.

The revolutionary German Renewable Energy Sources Act (EEG), now accepted by many of the German federal states as a best practice model, was pushed through in the face of this network of interests. In the year 2000, there were no professionally organized economic interest groups with a fixed position in government who lobbied on behalf of renewable energy and, compared to the conventional energy industry's active protection of interests, these groups are still weak today. The EEG would never have been passed if it weren't for the social movement towards renewable energy. Renewables had long been popular, sparking many municipal initiatives and finding political favour, primarily with the Social Democrats and Green party, and their parliamentary groups who formed the governing majority at the time the EEG was passed. The EEG was not a government initiative. Instead it was the creation of the parliamentary groups, supported by the majority of them and fought through against attempts to obstruct, even by their own government. It was the first energy Act to be passed in the face of massive resistance by the organized power industry. Its breeding ground was the social movement for renewable energy which had already made inroads into the political system. It was no longer the political influence of the established power companies which was decisive, but rather the legitimacy of renewable energy. This example shows that political parties and governments cannot only afford to be courageous when it comes to policies for energy change, but that this is increasingly expected of them. The most important political support for energy change is a population which is aware of the feasibility and advantages of renewable energy.

Increasing the number of players

The renewable energy movement is increasingly moving into areas in which recognition of its primary value coincides with self-interest. This has been my experience, gained from numerous speeches and discussions, whether with audiences drawn from a cross-section of society, or at conferences run by engineers, architects, boards of trade or industry, unions, manufacturing or agricultural associations, financial managers, young businesspeople, environmental associations or local energy initiatives. A wide range of players in society perceive a role for themselves in the fundamental shift towards renewable energy. It is only the die-hard "energy experts", fixated on traditional energy systems, who find this change problematic.

Massive public protests and resistance have recently and repeatedly prevented the construction of new coal-fired power plants, for example in Berlin-Köpenick, in Guben (Mecklenburg Pomerania), in Emden (Lower Saxony) and in Mainz (Rhineland-Palatinate). Claims that these power plants are still essential no longer have the desired effect. Instead, local campaigns argue for concrete alternatives involving each region's own renewable energy sources. Every time a conventional large-scale power plant is successfully prevented from being built, the transition to renewable energy accelerates.

Taking scientific studies as their basis, cities have begun to draw up solar registers to indicate suitable locations for private investment. They regularly conclude that solar power generators installed in inner-cities alone are sufficient to cover more than half the population's power needs. Municipal utilities, whether in small towns such as Wolfshagen in North Hesse or large conurbations such as Munich, are beginning to divert their investments towards renewable energy and want to win back their role as power producers. Scores of cities are buying back the electricity networks they had previously sold off, while in referendums the majority of citizens (e.g. 86 per cent in Leipzig) vote against the planned sale of their municipal utilities.

Industrial companies, especially those in the engineering and plant construction sector, discover they can use their existing know-how to become manufacturers of renewable energy technologies and to win a share of this future market. The number of companies running their own power plants to cover their electricity, heat and cooling requirements is increasing. They recognize the advantage of generating their own power – no longer must they pay network charges and nor must they continue to contribute to the profits of energy suppliers. All these steps towards energy autonomy involve renewables, even if they cannot yet cover 100 per cent of power needs. In June 2010, the German Telekom announced that it was making local smart-grids a key strategic project in order to "balance the network from below" – it would take too long to restructure a centralized system of energy supply in the same way. Thus the overall trend to decentralized initiatives will strengthen, decentralizing the use of renewable energy and emancipating businesses from the highly concentrated, and correspondingly rigid, power industry.

This process is also taking place in the credit industry. For decades, credit and investment activities in the energy sector focused on large-scale power plants, making financing the metier of major banks and large investment houses. With the renewable energy movement, and the financing to put it into practice, comes the recognition of just how advantageous investments into decentralized projects are: no large investment risks; the permanent avoidance of fuel costs; no installation delays; general adherence to budgets and short installation times which make capital returns faster. Creditors have grasped what energy economists either cannot, or will not, understand.

In a comparative study of the direct costs of generating power from renewable and traditional sources of energy, using the US as a basis and taking statistics from 2008, Thomas Dinwoodie has carefully calculated the actual differences. Assuming an operating lifespan of 20 years, a loan for 60 per cent of the investment sum at an interest rate of 7 per cent, plus a yield of 12 per cent, and uniform tax rates, he calculates costs of US$128 per watt for photovoltaics using crystalline cells and US$96 for thin-film cells, US$44 for wind power, US$74 for power produced from coal-fired power stations and US $98 for nuclear power.[60] And clearly these cost relationships will increasingly shift in favour of renewables over conventional sources of energy. This makes investments into renewable energy particularly interesting for insurance companies and pension funds, for whom long-term investments with no unforeseen risks are more important than short-term high yields.

These are all driving forces behind a growing and widening movement towards renewables, one which started out with individual solar collectors, photovoltaic systems and wind power plants. The first small steps in the movement already required the liberalization of regulations and laws in order to favour renewable energy. Until only recently, German public regulations had failed to consider renewable energy and thus energy change remained faced with numerous administrative hurdles, from municipal building regulations through to building law, regional planning laws, environmental laws, tax laws, land-use laws, mining laws and energy laws. In order to smooth the path for renewable energy in Germany, we must not only make changes to energy laws but also to the Federal Town and Country Planning Code (BauGB) which favours the authorization of conventional power plants, as well as Germany's Federal Nature Conservation Act (BNatSchG) which fails to differentiate between energy systems which do or do not emit pollutants. This process of political liberalization, which opens up the space in which renewable energy can develop freely and accelerates the speed of energy change, is still in the starting blocks.

The strategic and political key to accelerating energy change is to reverse the preferential treatment given to conventional energy suppliers and implicit in many laws. We must favour renewable energy for fundamental ecological, economic and social reasons and thus also for reasons of civil ethics. Both technological developments and society demand this change. The most important political task is to create a legal framework for renewable energy which society can then productively fill.

C. PRIORITIES: THE TIMELY ORDOLIBERAL FRAMEWORK FOR SOCIALLY ACCEPTABLE POWER SUPPLY

In order to accelerate the transition to renewable energy, this transition must be made a political priority and the primacy of renewables anchored in law. This involves more than simply putting a stop to the current privileges enjoyed by conventional energy, and it implies more than simply legal equality for renewable energy. Equality alone would not sufficiently reflect the fundamental differences between traditional and renewable forms of energy. Thus renewables must be given priority, over and above the legal dictates of energy law and in all spheres of legislation relevant to renewable energy. This requires fundamental political decisions, which must be permanently valid and based on a socio-cultural system of values. These will form the basis for a new, general legal framework that redefines public decision-making methods and responsibility and alters the evaluation and decision-making criteria of public bodies. These are vitally important structural framework decisions – in contrast to the sort of political decisions which simply make minor tweaks to problematic sectors and resemble piecemeal policies operating within a predetermined framework.

Once a new and universally applicable framework for developing renewables has been established, many individual political decisions will be superfluous. Key decisions designed to promote the transition to renewable energy make many laws and regulations on limitations to, and protection from, emissions increasingly superfluous, and wipe out the entire costs of controlling these emissions. These are "begin-of-pipe" regulations: with the comprehensive use of renewable energy, whose appropriation and use are pollutant-free, many of the damage reduction regulations, such as emission protection regulations, along the conventional energy flow or at its end ("end-of-pipe") fall away. Thus energy change also has the effect of reducing bureaucracy. Initially it will be hard to force through the fundamental priority for renewables, but this will simplify all subsequent decisions. The most important programmatic challenge is to recognize the key decisions that will open up a wide range of new developments. And the most important political resource needed to take these decisions is the courage to act.

Essentially there are four fundamental principles that need to be enshrined in law, reflecting both the physical energy imperative as well as social and ethical values:

- permanent priority for renewable energy in the power market;
- priority for renewable energy systems in regional policy and public land-use planning;
- the fundamental shift from taxes on energy to taxes on pollutants; and
- the stringent organization of energy infrastructure as a community asset, with municipal energy supply playing a central role.

These fundamental principles can only be implemented at national level, for they impact directly on each country's legal system. As a result, this development can only take place at a variety of speeds. In addition, now that we are forced to regard the efforts of the world climate conferences as having failed, for the reasons given in Chapter 2, we must decide upon the joint tasks for which the international community should be responsible. These questions will be dealt with in Chapter 6.

A definition of energetic/ethical ordoliberalism

Making the primacy of renewable energy a political norm is not a form of planned economy or economic statism. Indeed, nationalizing large power companies would be anything but a progressive step on the path to energy change. State-owned conventional power companies such as those in France, Italy, Greece or Austria, whether in the electricity, oil or gas sector, have proven every bit as negative in their attitude to renewable energy as privately-run power companies. State-owned power companies also have a direct influence on their national governments, and these can have no interest in rapid energy change if this renders their own revenue-generating power companies uneconomic. To nationalize power companies which own nuclear and coal-fired power stations would be like creating a public "bad bank", in which all the bad risks were bundled together. Thus prioritizing renewable energy does not mean greater state involvement in energy production.

If nothing else, the statutory primacy of renewable energy constitutes a regulatory market framework, one which is not arbitrary but instead is legitimized by its goal of protecting public welfare. This is the central tenet of ordoliberalism, and the exact opposite of neo-liberalism, whose social "damage potential" has been proven in recent years. An ordoliberal economic approach sets standards applicable to all businesses while avoiding micro-economic political interventions unless there is a compelling systemic interest in doing so. In particular, it reflects the indispensability of public infrastructure which must be equally accessible to all economic participants, producers as well as consumers, and under the same conditions, thus ensuring the economic principle of competitive and consumer equality. This is what Jan Tinbergen, winner of the first-ever Nobel Prize for economics in 1969, calls "social overhead capital". This was a basic principle common to all economic theories before neo-liberalism levered it out and forced public infrastructures to meet target yields. Classic features of ordoliberalism are the prevention of economic monopolies, the principle of competitive equality, and social obligations which must be met by all economic participants equally – principles which lead to a social economy. That these social obligations must also be ecological obligations is now evident, thanks to the high social cost of using environmentally damaging resources. And for just this reason, the statutory primacy awarded renewable energy must be permanent. An economic order in which polluted water has the same market value as clean drinking water is an indication of social neglect.

The general, statutory primacy of renewable energy would have a more resounding effect than all other political approaches and thus will need to be enforced in the face of massive resistance. Yet with resolute action this is easier to achieve, for it is a simpler message to convey to the public. It offers greater transparency and equality and enjoys a decisive psychological advantage: a major political step such as this would meet the growing social need for a big solution and a quick way out of the current energy trap. It achieves what large-scale technical projects only promise. Political initiatives for energy change, which one knows from the outset cannot be fulfilled because of the scale of the challenges, are not enough to shake people out of their lethargy. By giving renewable energy social priority we provide a new basis for energy supply. This activates innumerable social initiatives for investment into renewable energy and encourages joint responsibility.

1. Priority for renewable energy in the power market

The most prominent international example of the politically safeguarded primacy of renewable energy is the German Renewable Energy Sources Act (EEG), whose surprising success is proof of the rapid effect of such a statutory principle. The regulations in the EEG which determine the primacy of renewable energy are based on three fundamental ideas:

- The first is that all green electricity has *priority network access* over conventional power; thus green electricity cannot be refused entry to the public power supply network with the argument that network capacity is already exhausted by conventional power plants.
- The second element is the *guaranteed feed-in tariffs*, set at a level which covers investment costs and allows for yields. The tariff levels vary between the different sources of renewable energy, for these are at differing stages of development and cost levels and should be encouraged to develop further. The plan is to systematically increase the share played by renewable energy in overall power supply, while simultaneously reducing the role of conventional energy. The level of feed-in tariffs is politically determined. The tariffs are only paid to the producers of green electricity, not to power plant manufacturers. From time to time these tariffs will be reduced as costs continue to fall, yet only for new plants as only these have the degression advantage of increasing plant productivity. Feed-in tariffs apply for only a limited period. This second element creates *investor autonomy* for renewable energy: operators of renewable plants no longer need to ask conventional power companies whether the investment fits in with their own capacity planning or not. While the guaranteed feed-in

tariffs generate additional costs, these will be equally shared between all electricity customers, although with a few exceptions for companies with high electricity consumption levels, for example the aluminium industry. The principle of feed-in tariffs, which the Act's opponents denounce as running against the principles of a free market economy, actually has the effect of generating tough market-oriented competition in productivity among power plant manufacturers. The guaranteed feed-in tariffs function as an incentive for purchasers of these plants to order the most efficient and productive systems, for this reduces their costs and increases their yields. No other political policy for introducing renewable energy has contributed more and faster to an increase in power plant productivity and given greater impetus to the global renewable energy industry.

- The third element in the Act is that there is *no feed-in limit* for green electricity. Where renewable energy is introduced only to a limited extent and market expectations are low, power plant manufacturers hold back. They need the perspective of a steadily growing market. By opening up this perspective, the German EEG has become the trigger for the global industrialization of renewable energy technologies.

Market regulations for meeting public objectives

This preferential Act is a form of market regulation based on several undoubtedly important public goals: not only that of CO_2 reduction, for reasons of both climate and environmental protection, but also long-term energy security, achieved by ending energy imports and supporting regional economic structures and going hand-in-hand with increasingly decentralized energy supply. It is a *parallel energy market regulation*, in addition to the standard one. And that means: if allowed to continue unrestricted, this Act would remain valid until conventional power supplies had been completely phased out, and then itself become redundant. It would lead us to an energy market based entirely on renewable energy, one subject to market rules of which we are currently unaware. These rules will be strongly dependent on the extent to which power production can be autonomous, the relationship between individual, local and regional suppliers, and the differences in cost between the various forms of renewable energy. We will be dealing with a different form of "power culture", one which develops on the path to renewable energy.

Laws on feed-in tariffs for renewable energy must continue to be developed in order to meet the new challenges and opportunities which arise. However, there must be no discontinuation of the principle of the primacy of renewable energy, neither when price equality with conventional energy has been achieved, nor when it becomes even less expensive than its conventional competitor. Those who hurry to suggest just this misunderstand the nature of the EEG, seeing it purely as a price regulation mechanism. Although from a certain point onwards the cost and price advantages of renewable energy mean that guaranteed feed-in tariffs can be dispensed with, the same does not apply to the primacy enjoyed by renewable energy which continues to enjoy greater social value over conventional energy. The key to continuing the momentum of energy change as introduced by the EEG is the strategic question of which steps must then follow.

As the role played by green electricity in the existing power supply system grows, so the systemic conflict between renewable energy and conventional energy will come to a head. Green electricity, fed into the network as a result of the purchase guarantee, cannot be regulated in the same way as the power it replaces. The periods in which more solar and wind-generated power flows into the network than is actually needed will become more frequent and longer. The challenge facing a system of renewables-based power supply is not whether supply is sufficient, but what happens with the intermittent excesses of power, generated during periods of high solar radiation and strong winds, that is not immediately used. It would make no sense to stop wind power and solar systems from producing power, for the production process involves no fuel or processing costs. The only satisfactory answer to this question is to store the excess power.

Continuing to develop the EEG

Now we need systemic incentives which are no longer focused at integrating renewable energy into the conventional power market but on adapting this market to the needs of renewable energy. This must be the goal of the EEG as it continues to be developed. Just as the EEG has become an international role model, so will its future form serve as a model for all those countries in which the role played by renewable energy has reached system-relevant dimensions.

Therefore the most important next step is to introduce a combined *power plant bonus* as outlined in the concept designed by Jürgen Schmid of the Fraunhofer Institute for Wind Energy and Energy System Technology in Kassel. According to this concept, feed-in tariff levels would vary according to three periods during the day – one period being subject to a minimum tariff and the two others to higher tariffs, enabling investments to be made into storage capacities. During two of these periods, in which production levels reflect actual radiation or wind conditions, a portion of the generated power is stored before being fed into the network during the third period. Which hours these periods actually cover can be agreed with the network operator 24 hours in advance. This idea would ensure that power is fed into the network when needed, and would encourage a wide range of investments into storage mechanisms.

A second step will be for network operators to issue tenders for *capacity quantities*, delivery of which is then guaranteed. This means the guaranteed supply of an agreed quantity of power within an agreed period of time. This too would be a means of adapting the power supply system to renewable energy because these capacity quantities would need to reflect actual power demand. A third step would be to set the criteria for national regulatory authorities' so-called incentive regulations so that they favour investments in *smart grid structures*. The German regulatory authority, the Federal Network Agency (BNetzA), approves the prices which network operators are permitted to set for transmitting and distributing power to local networks. They can also prevent network investments or their write-down. The BNetzA acts in this way because its current primary concern is to lower transmission and distribution prices, although this has the effect of putting the brakes on investment into adapting the network for renewable energy and the corresponding network management. This method of directing investment according to very one-sided criteria is counter-productive and serves to preserve the current structure. Therefore, in order to secure the priority of renewable energy, there must be a legal requirement that all network investments for renewable energy be recognized and permitted to be reflected in prices.

In addition, operators of renewable energy power plants must have the legal right to construct their own connection lines to the power network, and to charge the same transmission and distribution prices as network operators. This is necessary in order to prevent network operators delaying expansion of the power supply network.

2. *Priority in urban and land-use planning*

The most important factor in accelerating the role of renewable energy is to eliminate any arbitrary policies which hinder site authorizations. The braking effect these have has been demonstrated in Chapter 1. By refusing site authorizations, even the best market regulations for renewable energy can be actively undermined. Ending this strategy of hindrance is extremely important. In the end, the way in which this problem of site authorization is solved will be decided by developing renewable energy structures, whether this development lies in the hands of many or just a few, and whether the opportunities for industrial renewal and wide economic value creation can be used.

Construction regulations are set out in planning and land-use laws. Only in the latest amendment to the German Regional Planning Act (ROG), passed in December 2008 and as the result of initiatives stretching back years, were climate protection and renewable energy explicitly considered. The ROG determines that regional land-use must accommodate climate protection considerations and create the preconditions for the expansion of renewable energy. However, because of Germany's division of competencies between Federal and State governments, as laid down in its constitution, the Act's legal implementation relies on adoption by the latter, and this has, to date, been either inadequate or totally lacking. The ROG designates a geographical area's social functions as being of public interest, and construction that reflects this public interest enjoys priority when it comes to authorization. Where there are conflicting areas of public interest, these must be evaluated by the authorities and, where necessary, decided upon in court, often in long, drawn out procedures. Public interests include regional transport routes such as roads and rail tracks, energy networks, the designation of industrial land-use areas, nature conservation, agricultural subsidies and regional development focuses. Added to these are water and environmental laws with their own exemption criteria.

An authority which wishes to deny or delay site permission for a renewable energy power plant can almost always find a reason to do so among the multitude of applicable regulations. A popular method for authorities to avoid accusations of discrimination (which become obvious where refusal is total) is to designate just a few priority areas and to declare remaining land an exclusion area, making widespread and rapid energy change practically impossible. In states such as Baden-Württemberg, Bavaria and Hessen, this method has been used to prevent the installation of wind power plants on 99.8 per cent of available land. This is effectively a prohibition on investment into wind power and makes it almost impossible to use the potential offered by small-scale hydroelectric power plants. However, in planning laws to date, renewable energy has not been recognized as a public interest. In contrast, electricity transmission networks, for example, enjoy statutory planning privileges which must be strictly complied with.

As a decentralized power supply structure based on renewable energy will effectively replace a limited number of large-scale conventional power plants with a multitude of widely scattered plants, this requires a multitude of authorizations which, in turn, offers many opportunities to block this development. It is impossible to calculate just how many renewable energy investments have been foiled to date, thanks to obstructive planning and long-winded authorization procedures. It must be many thousands in Germany alone. Authorities frequently demand expensive studies which must be financed by the applicants themselves, without it being any more likely that permission will be granted later.

The second decisive step is to establish priority for renewable energy in planning law, for all aspects and stages of public land-use planning, in addition to its priority in the power market. Renewable energy must not only be legally codified as a public interest, but must also enjoy priority over competing public interests. This can be justified in that renewable energy simultaneously fulfils a whole range of criteria which are defined as public interests. It meets the demands for nature conservation and environmental protection, for these can only be guaranteed by the comprehensive transition to pollutant-free energy use. If the laying of power lines is regarded as public interest for guaranteeing supply security, then this must also be the case for locations in which power is generated from renewable energy. Although renewable energy is generated regionally, its importance is undoubtedly more wide-ranging. It meets all the criteria of regional industrial and business development as well as agricultural development in rural areas, over and above the potential for the production of energy crops – for agricultural areas can serve as sites for wind power plants without losing the ability to be simultaneously used for agricultural purposes.

When the provision of renewable energy is declared a *priority public interest* in regional land-use policy, then this leads to a fundamental re-weighting of administrative and legal procedures. When the various public interests are compared, arguments for the priority of renewable energy are universally plausible: compared to other environmental and land-use interests, the benefits of renewable energy for the environment and landscape are clearly greater and more comprehensive. Renewable energy plants may be intrusions in the landscape, but they make an indispensable contribution to protecting the basis of our existence, i.e. nature and the environment. Forbidding these intrusions during the general transition to renewable energy, by refusing to authorize sites, necessitates increased numbers of intrusions in other regions, including the greater impact on nature and the environment made by energy transport networks. All regions can contribute to the process of energy change and, for reasons of fairness, all should play their part. The politicians who are unable to explain this clearly to the public have either failed to understand the concept themselves, or are too cowardly to face down short-sighted and self-serving objections.

The fundamental decision to elevate renewable energy from its position of effective subordination and to make it a priority in public land-use planning is a move from the *passive to active conservation of nature*. Passive nature conservation focuses on preventing interventions in the natural environment, with the consequence that the interventions which do occur become more spatially concentrated. In contrast, active conservation aims at harmonizing human activity and the natural environment as far as possible: the solution to the challenge of unifying business with environmental protection is to integrate them at regional and local level, rather than continuing to keep them apart.

The basic outline for statutory priority in public land-use planning

I developed the first basic outlines for such a priority act in 2008. It became the draft law for the Hessian parliamentary group of Social Democrats (SPD), during the time at which its leader, Andrea Ypsilanti, was standing for the post of Hesse's Minister President.[61] She wanted to pass this draft law in order to stimulate a "spring-tide of new investment in renewable energy", as she explained. Supplemented by a 100,000 mini-cogeneration plant programme, which alone could replace two large-scale 1,000-megawatt power plants, this law would have been able to boost the contribution made by renewables to Hesse's power production from 5 to 60 per cent within only five years. However, this draft law was never passed because of obstruction by four rebel Social Democrat parliamentarians who prevented an SPD government taking over in Hesse. The trailblazing role of this draft law, which aimed at definitively introducing energy change and achieving 100 per cent renewables-based power supply by 2025, had been understood and led to strongly polarized conflict. If this draft law were to be passed, commented the national German broadsheet, the *Frankfurter Allgemeine Zeitung*, then "those wishing to prevent investment in renewable energy would have no more opportunity to do so". Even so, campaigns against this political undertaking were unable to prevent its growing popularity.

Although this draft law defined areas in which wind power plants would enjoy priority, it prevented any further exclusion areas being established (with the exception of nature reserves). Wind power plants in priority areas would generally be authorized. Authorizations for installations outside priority areas would be decided upon solely by the municipality upon whose land the power plant would be installed. But no municipality may declare itself an exclusion area. There are no longer across-the-board plant height or size restrictions – all decisions must be taken on an individual basis. Furthermore, municipalities have the right, by means of statutes, to prohibit the use of particular fuels in all or part of the municipality, and to stipulate the active use of particular forms of renewable energy, especially for buildings and other structural installations. This framework indirectly obliges every municipality to actively drive forward, and take on joint responsibility for, energy change. At the same time it provides each municipality with the independence to make its own decisions, effectively democratizing energy change. To ensure that municipal decision-makers fulfil their duty, and do not react arbitrarily to investment initiatives, the responsibility they have been given will lead them to define their own municipal priority areas and include the active use of renewable energy in their own development plans, if only as a means of supporting their local economy and boosting municipal tax revenue. Together with their public utilities, municipalities will also be encouraged to take back control of power supply. The concrete authorization of new plants must be made in accordance with the zoning plans authorized by municipal councils.

All this serves to direct the active use of renewable energy at the level of the greatest number of potential activists, with the greatest economic self-interest as well as the best and most efficient opportunities for using a combination of available renewable energy sources at regional level and for a variety of uses. This includes extracting the energy bound up in organic waste, i.e. the integration of waste management and energy supply which alone could cover up to 20 per cent of power and heating needs; the use and processing of agricultural waste and energy crops as well as recirculation of the waste matter resulting from biogenic energy production (e.g. degassed biomass from biogas plants, oilcake from vegetable oil production, residue from the production of bioethanol) as animal feed or fertilizer on farms around the municipality; the combined production of power, heat and cooling in tri-functional cogeneration plants; making use of the energy inherent in surface geothermal energy which could completely cover the power and heat demands of buildings; power and heat network management by public utilities drawing on the surplus production and reserve capacities of independent private operators with combined power and heat tariffs, something largely possible only with decentralized structures and which gives public utilities a natural competitive advantage. A shared feature of all these activities is that they largely avoid transport costs. By its very nature, this regulatory policy works best at municipal level. This is the only way of utilizing all the opportunities for the efficient use of renewable energy.

This was also confirmed by the new EU building code passed in March 2010: it directs that all new public buildings built after 2012 must meet zero-emissions standards, that all building renovations must meet statutory minimum standards for the use of renewable energy by 2015, and from 2020 for the construction of all new buildings, without exception – a standard which can only be met using local sources of renewable energy.

The potential for using areas creatively

An example of the magnitude of the potential offered by using areas creatively was given in a memorandum published by EUROSOLAR. It sees Germany's longest autobahn, the A7, which runs 960km through Germany from north to south, as an "energy alley" and suggests lining all the stretches of the autobahn suitable for appropriating wind energy (roughly 80 per cent of the entire length) with 5-megawatt wind power plants, around 780–900m apart.[62] This would give a total of 1,250 individual sites and a total installed capacity of 6,250 megawatts, which alone could cover 2.2 per cent of Germany's power needs and would require investment of €7.5 billion. According to their precise location, the blades would rotate around a hub at a height of 100–130m. The systems would be installed in areas which are already intensively used. By declaring the project a priority, all the authorization procedures would be swept away, ensuring a sufficient number of willing investors. Similar installations could be made along other German autobahns. In order to cover 15 per cent of Germany's total power needs (the official target of the DESERTEC project), using the same criteria as for the A7 and with the current costs for 5-megawatt wind power plants, a total investment of €50 billion would be needed. Investments for connections to the public grid would be relatively low, as many power lines already run parallel to autobahns.

Most importantly, this project could be begun immediately and would require little more than three years to complete. Power generated in this manner could be used primarily to recharge electric cars at filling stations along the autobahn. Wind power alleys can also be constructed along railway tracks, especially along main and high-speed lines, with the additional advantage that the power generated can be directly used for the locomotives. This method of designating priority areas is also relevant for solar power production: for solar modules at the sides of autobahns, mounted on the noise barriers or along stretches where autobahns are partially covered. For autobahns passing close to built-up areas, or four-lane main roads running straight through towns, these would be urban land-use design projects. There is no need to move onto agricultural land in order to increase the available areas.

Once power-generation facilities based on renewable energy are expanded outside the priority areas, with wind power simultaneously becoming an integral component of municipal land-use planning, and solar generation of building design, then the decentralization of power supply structures will continue inexorably. Integrated, consumer-friendly forms of production open up functional synergies, and these make redundant any calculations of energy efficiency based on single, individual installations. Energy supply outside priority areas will no longer be spatially concentrated, but rather will become an integral part of future rural and urban landscapes, with widespread production, little need for transmission, and reduced transport costs. Taken together, this underlines the far-reaching consequences of making renewable energy a priority in urban and land-use planning.

The extent to which energy change is currently limited by arbitrary planning restrictions is also demonstrated by the extent to which "repowering" wind power plants lags far behind the possible and the expected. Repowering means replacing a low-capacity wind power plant with a high-capacity one. This assumes that authorizations for taller plants will be granted. However, because these authorizations are overwhelmingly refused, the absurd result is a deliberate political prevention of more cost-efficient wind power generation. This proves that the most important task is to exhaust existing political potential.

Unless priority is given to specific locations, the opportunities for decentralized production will be arbitrarily limited or will remain stuck fast in the bureaucratic maze of interminable and expensive authorization procedures. It is an intolerable form of pre-democratic paternalism, harking back to an era of deference to the establishment, in which government authorities continue to assume greater powers of decision-making over site conditions than municipal decision-makers with their knowledge of, and democratic accountability to, their local communities.

3. *Pollutant tax instead of energy tax*

The most important approach for implementing the principle of priority for renewable energy would be to transform energy taxes into a tax on pollutants. This idea is as unconventional as was that of priority for renewable energy in the power market, something completely unheard of a quarter of a century ago. The "eco-tax" everyone had been talking about (but of which much less is now heard) implies at best a move in the direction of a tax on pollutants. The same goes for the carbon tax which is the subject of wide discussion and already levied in several countries. Eco-taxes and carbon taxes supplement existing energy taxes but do not represent a fundamental restructuring of the system of energy taxation. The key idea is that all energy taxes, including the eco-tax, be replaced by a pollutant tax. The level of this pollutant tax must reflect the actual quantity of pollutants being emitted and must be sufficiently sophisticated in order to do so.

A pollutant tax would trigger long-term change as well as alterations to patterns of production and consumption. This is also the intention of eco-taxes, which stopped being developed years ago and whose effects have not been wide or deep enough. The psychological challenge inherent in the term "eco-tax" is that every tax is seen as a burden, although it actually serves to reduce the burden on environment. This is just one reason that eco-taxes have found it difficult to win public acceptance. Another is that eco-tax revenues are not directly linked to ecological investments. However, when every form of energy is taxed according to its actual polluting effect (CO_2 emissions, dangers to health, atomic waste, water pollution, etc.), then this leads to minimal or even zero taxation of energy forms with few or no pollutants. At the same time, taxing energy sources that contain pollutants or cause damage compensates for the real subsidies they currently enjoy in the form of failing to pay for the environmental destruction they cause. A tax on pollutants encourages both producers and consumers to change to low-pollutant and pollutant-free forms of energy. It is *the* instrument for avoiding social costs and encouraging the widespread technological developments necessary to do so. The indubitable economic benefits of an environmentally- and resource-friendly economy will thus be achieved through a microeconomic approach.

The precondition for introducing a tax on pollutants is to develop a "pollutant formula" which, based on the multifaceted environmental impact of fossil fuels and nuclear power, gives the tax a scientific basis and allows it to be clearly understood by the public. Campaigns against a pollutant tax, similar to those being run against eco-taxes, will be faced with an almost insuperable psychological obstacle: anyone demanding that the pollutant tax be lowered is publicly admitting their willingness to further burden society with pollutants. This attitude is no longer socially acceptable. A pollutant tax is also a more effective means of climate protection than a pure carbon emissions tax – a carbon emissions tax only compensates for the numerous negative environmental aspects of fossil fuels when these occur synchronously with CO_2 emissions. Despite using fossil fuels, CCS power plants are not affected by the carbon tax. But neither they nor nuclear power plants would escape a pollutant tax.

This fundamental shift from an energy to a pollutant tax naturally raises the question of future state revenues, a significant share of which currently come from taxes on energy. The more the pollutant tax starts to bite, and the more producers and consumers shift to pollutant-free sources of energy, the smaller the revenue from this tax inevitably becomes. However, when the overall fiscal situation is taken into consideration, there is no danger of tax revenue drying up: when a pollutant tax is introduced, the initial revenue from the pollutant tax would roughly equal that of current energy taxes. This tax revenue only starts to fall away, step-by-step, once investments have been made in pollutant-free renewable energy – and these investments represent additional economic activity which itself has a significant dynamic effect, generating new jobs and boosting business. The result is an increase in tax revenue from other sources, one which more than compensates for the drop in pollutant tax revenue. As a consequence, not only will national economies flourish, but so will their living standards and environmental quality, while social costs will fall.

Naturally, this concept is only practical in countries that already have an energy tax. It is hardly applicable for developing countries in which energy consumption itself is often subsidized, with the purchasing power of the local population insufficient even to buy untaxed energy. The concept would, however, be applicable for oil and gas producing countries whose enormous state revenues mean they have no need of energy taxes and where energy prices are kept extremely low. Here we would not be looking at a change to the tax system, but rather at a charge with which to awaken the public's awareness of the need to protect their environment and to use energy sparingly.

D. PUBLIC PROPERTY: THE KEY ROLE OF MUNICIPAL POWER SUPPLY

No one disputes that economic monopolies are misused to create market barriers to other, often better and more productive, commodities. This occurs irrespective of whether monopolies are nationalized or privately owned; to date both state and private monopolies have worked in a similar manner to stem the tide of renewable energy. Thus liberalization of the power markets, long a feature of the petroleum industry and introduced by many countries in the power and gas sector during the 1980s, has not worsened the conditions for renewable energy. To some extent it has even made minimal improvements, by opening the market to producers of green electricity. On the other hand, this liberalization has also increased the mental barriers to renewable energy because, as described in the section on market autism (p71), renewables are being required to establish themselves in a market skewed to favour the status quo. Moreover, these changes have been made only half-heartedly, which is especially clear in the electricity sector. The key to effective liberalization is to prevent individual companies having control over the production, transmission and distribution of electricity. However, at the corporate level there has been no requirement to separate ownership relations in such a way.

An electricity producer who is simultaneously the owner of a power transmission network is not permitted to award itself preferential transmission and distribution tariffs and to discriminate against its competitors. However, a network is a natural monopoly. It must be used by everyone and, for spatial and functional reasons, competing networks are impossible. This is the case for all earthbound road, rail, communications, water, gas and electricity networks, and therefore they must each be required to operate in a neutral manner with respect to all their users. This can only truly be guaranteed when networks are publicly owned and not required to generate yields, because it is in the general public interest that even uneconomic sections of the network be maintained. If we needed to deal only with non-discriminatory tariffs for network use, then a regulatory authority would be sufficient.

But there are two other key aspects to electricity networks. On the one hand, they have a control function, because the commodity they are transporting and distributing is not a physical entity and feed-in and uptake quantities must be synchronized. Thus the network operator decides upon the quantities and source of the electricity fed in to the network. And on the other, a network operator must design the network according to the production locations. However, this means that it is entirely up to the network operator whether, and how quickly, it adapts to the structural changes associated with the transition to renewable energy, from a few large-scale power plants to many decentralized electricity generators, and whether the operator is prepared to connect these new generators to the electricity network. Where operators delay or refuse to do so, then they retard or inhibit this structural transformation.

This has clear consequences for the goal of rapid energy change. The natural monopoly enjoyed by electricity networks must be held in the public hand and democratically controlled. This is particularly the case for municipal networks that run at low, as well as medium, voltage levels, and to which the vast majority of decentralized electricity producers are connected. (Here connections between decentralized producers and the network can be established fastest, for the transmission lines needed are usually short.) This is yet further proof that the fastest route to energy change is through decentralized power generation; the processes are manageable and municipalities have a greater interest in decentralizing electricity production than do centralized network operators or power producers – all arguments for municipally-owned networks. Municipalities who sell off local networks slow down the process of energy change – bringing networks back into municipality ownership accelerates it.

Transmission networks, too, must be socialized in a democratic manner: not only must they become public property, they must also be subjected to efficient, democratic control. This is the case in Denmark and Sweden, for example, where transmission networks are exclusively publicly owned. Existing networks cannot be permitted to be a playground for corporations who, listed on the stock exchange and required to generate yields for their private shareholders, have no interest in making new investments at the expense of the transmission tariffs permitted by regulatory bodies. A publicly operated power and gas network needs only to cover its costs, and regular auditing will ensure operational efficiency. All the reasons for arguing against the privatization of the railway network can be used to argue that the transmissions network should also be in public hands – the publicly owned "natural monopoly" of network structures for the optimal use of publicly owned renewable energy. Thus the German government is guilty of a major political omission by not purchasing the transmission networks sold in 2009/2010 by the two German power companies E.ON and Vattenfall and establishing a publicly owned network company. Not even the two political parties in Germany who support energy change, the Social Democrats and the Greens, pushed hard for a public takeover. There was no wide public debate similar to that over the privatization of the railway network, as if the transmission network were of lesser public importance.

Infrastructure synergies

There is a further reason for public network ownership. It relates to the overall future design of all earthbound infrastructures which, to date, have been independently constructed and operated: the electricity network, road network, railway network, waterways and the water supply network all lay claim to, and shape, the landscape. A system of power supply based on the technological diversity inherent in renewable energy makes synergies both possible and socially productive: main roads with integrated power lines, in the form of either underground cables running along the middle or at the edge of autobahns; water supply networks integrated into hydroelectric plants and pump storage power plants. Synergetic solutions such as this are economically productive and reduce the burden on the landscape. The ideal state of affairs would be a public network operating company for all these infrastructure networks, subject to a controlling body involving not only state representatives, in their role as public owners, but also independent representatives of public interests – municipal associations, environmental associations, consumers associations, industry associations and unions. Admittedly, this could lead to the problem of a static organization of mammoth dimensions. Therefore it would be advisable to realize synergies via a joint *network operating agency*, which issues an annual public infrastructure report and recommends concrete decisions on synergetic infrastructure investments.

Citizen value

There is no danger of oversized organizations at municipal level – here infrastructure synergies are of direct, practical relevance. This supports calls for a renaissance of the traditional municipal utility, once again taking up its role of network operator, running its own power plants with power and heat storage capacities and power or biogas filling stations, as well as water supply, waste management and processing, and local road construction. As Munich's Mayor Christian Ude so aptly stated during his speech at the EUROSOLAR conference on "Renewable Energy For Municipal Utilities", municipal utilities are not bound by any criteria of shareholder value, rather they must operate according to criteria of "citizen value". To privatize municipal utilities is to be blind to the future and undermines the opportunities and needs of public services in general. Bringing back privatized public utilities and networks into communal ownership, or the setting up of new public utilities, is the fundamental precondition for rapid energy change, for the productive use of network synergies and for a generally more productive supply structure. They help to win back decision-making powers for municipal self-management and act as a stimulus to municipal democracy.

Modern power supply began with the public utilities. As power supply became centralized, so public utilities became marginalized, often abandoned and sold off. But they represent the part of the traditional power industry which needs to take on the dominant role in a power supply system based on renewables. They will take on an importance far greater than their original role, becoming the key supporters of publicly owned infrastructures, without which there can be no socially acceptable economy. They can overcome the "Tragedy of Commons" spoken of by Elinor Ostrom, who was awarded the 2009 Nobel Prize in Economics for her theories on the concept of public property. Municipalities can again become supporters of "commons law" by making renewable energy, a public property, locally available. Thus they represent an alternative to the endless globalization of existential, basic social needs.[63]

Where municipalities are awarded the right to authorize site locations for new forms of energy production, they also have the opportunity to create social balance: every authorization allowing a privately-owned wind power plant to operate on open land gives the owner of that land a financial privilege not shared by others – as is the case in building legislation, where the landowner benefits when his property is declared building land. This generates envy and social imbalance. As a counterbalance, municipalities should switch to giving priority authorization to cooperative operators, or give public utilities the pre-emptive right to purchase or lease the site. This is in keeping with the public nature of energy supply.

In his thesis on the role of municipalities in local power supply, lawyer Fabio Longo convincingly argues that renewable energy initiatives are classic examples of local political duties as defined by the German constitution and other current laws. In Germany's Basic Law, municipalities are guaranteed the right to regulate all local affairs (Article 28, Paragraph 2), including the ecological aspects of settlement areas, such as local air pollution control, local climate protection and use of the natural local energy cycle as it relates to construction matters. The municipalities are also given political responsibility by the German Federal Building Code (BauGB, Chapter 1, Part 6) to provide local public infrastructure, and it sets out their responsibility for urban land-use planning as well as efforts to create a "medium-sized infrastructure for the purposes of consumer-oriented public supply" which can only be guaranteed by the local – and thus decentralized – supply of renewable energy.[64]

Incidentally, the municipalities are not themselves required to act as investors. They may limit their activities to creating the space for private investment and investment by public utilities. Even where investment is private, it remains decentralized – value creation is local, installation and maintenance work is carried out by the local workforce – and is not limited to the few locations controlled by centralized power suppliers.

Decentralization in this sense serves to balance society's living standards. Imagine a region with a population of 1 million, all of whom are currently supplied with energy by centralized providers. In Germany, current per capita energy costs (excluding plant investment) are around €2,500 per annum. This includes all direct and indirect energy costs, i.e. power, heat and fuel, as well as the energy costs represented in every consumer commodity and service. This represents a total of €2.5 billion per annum flowing out of this region's local economy. With the complete transition to power supply based on local renewables, this €2.5 billion would remain in the local economy. This is the equivalent of an economic development programme of the same magnitude – on an annual basis, with no bureaucratic effort and distributed amongst the whole population! No government could ever afford to fund a development programme of this size. It says much about the energy economists who, focused on energy-efficiency irrespective of its source and continuing to stare exclusively at kilowatt hour prices, fail to see this connection.

5

PRODUCTIVE FANTASY

Energy change as an economic imperative

As our power supply system is based on conventional energy, the entire business management sector sees the energy problem only from the perspective of the conventional power industry. What the conventional power industry defines as efficient, cost-effective or feasible has become the generally accepted definition of the terms efficiency, cost effectiveness and feasibility – and not just within the energy industry itself. The traditional power industry's systemic economic terms have become general economic terms and its claim to omnicompetence part of its self-image. When the environmental movement of the 1970s made nuclear and fossil power supply a social issue for the first time, the traditional power industry reacted by demanding a "return to energy consensus", by which they meant taking the politics out of energy. "Energy consensus" involved prohibiting political intervention and making all forms of criticism taboo. For decades both political institutions and "the economy" have respected, and accepted, the power industry's self-appointed position of authority. This has led to society's division of roles between energy providers and energy consumers, and an economic division of labour between energy suppliers on the one hand and the producers of commodities and services on the other, divisions which exist right through to the present day.

As a result, it is difficult for many companies to perceive the opportunities that energy change opens up to them. Instead of staring, transfixed, at constantly increasing energy prices, they ought to be considering their own future role as producers and users. Renewables-based power generation relies primarily on new technologies; it has increasingly less – and sometimes nothing at all – to do with energy distribution. Only once this has been understood does it become clear that renewable energy is of economic importance to all sectors of the economy, and it is the technology companies themselves who can, and must, take the leading role. The development and production of renewable energy technologies will spark off a new economic revival. Yet in contrast to past boom times, the boom initiated by energy change will bring about a fundamental qualitative change. The real economy will no longer have to deal with the limits and consequences of conventional energy growth. When decentralization has made regional resource cycles a matter of course, no longer can anyone be deprived of the fundamental commodity – energy.

Renewable energy differentiates the debate on growth. Economic growth will be linked to environmental protection and natural growth – and thus to Earth's only true growth process: that which is driven by the sun. Biosynthesis, i.e. the growth of vegetation, turns entropy (the law of the degradation of energy) into negative entropy; pollutants (at least in power supply, although not necessarily in the processing of other industrial raw materials) are converted into productive growth. Independent calculations of increased profits, achieved at the cost of the natural environment (i.e. the increased damage and destruction of resources) then truly become added economic and social value. The contradiction between ecology and economy is eradicated; the more so as economy can simply become a subcategory of ecology. Ecology describes the environment available to us. This means the economy, in its true sense, its efficient and provident use. The true contradiction we face here is that between an ecologically-damaging, and thus ruinous, economy, and an environmentally-friendly economy that benefits both our natural surroundings and society.

A. SYNERGIES: NEW PRODUCTS FOR MULTIFUNCTIONAL APPLICATIONS

The manufacturing economy also stands at this fork in the road. The more that companies actively position themselves to take advantage of the opportunities offered by energy change, adjusting their production accordingly, the greater the number of new fields of activity available to them. Energy change opens up new growth opportunities in several traditional industrial sectors. The electrotechnical and metal-working industries will maintain their leading role with renewable energy. Other major industrial sectors will be faced with new opportunities to develop and diversify if they utilize the synergies generated by renewable energy. However, several industrial sectors will need to completely change the focus of their activities if they wish to cast off their role as obstacles to progress.

A change of engine and energy for the automobile

An example of this is the automotive industry. It is in its own interest to free itself from its long and unholy alliance with the petroleum industry. There is no long-term market future for its current product range of vehicles with combustion engines. Instead the industry must offer vehicles with drive systems based on alternatives to fossil fuels. If it fails or delays in taking this step, it will be faced with a shrinking market because, with the approaching exhaustion of fossil energy sources, fuel prices will unavoidably increase, making the costs of running a car problematic for many of the industry's customers. Thus the general trend towards electric vehicles is pre-programmed; they offer not only the greatest levels of energy efficiency but also the lowest running costs. At around €1.4 per litre of fuel, even a 5-litre car has running costs of €7 for 100km; an electric car using today's technology has running costs which are only a third of this amount. This alone guarantees that demand for electric cars will increase in leaps and bounds.

In the meantime, the automotive industry has recognized that the electricity used to drive electric vehicles will, and must, be generated by renewables. It is not even necessary to forcibly link the introduction of electric cars with their use of green electricity – this would be almost impossible to control and would act as a brake on development if the mass introduction of electric cars were made dependent upon the transition to renewable energy having already been completed. This would, in effect, be an embargo on the introduction of electric cars, an embargo which would be almost impossible to implement politically. The time is nearing in which small and medium-sized electric cars with a range of between 200km and 300km will be available at prices comparable to conventional vehicles.

It is sensible to make the transition to electric vehicles a political priority because their mass production will lead to the development of ever more powerful and cheaper batteries, making the electric car the driving belt of energy change. The ideal political method of speeding up their introduction is a funding programme, perhaps in the form of low-interest loans. In order to simultaneously drive forward the production of green electricity, this funding should be made contingent upon automobile manufacturers investing into the additional renewables-based generation of electricity at a scale equal to the average electricity consumption of the electric cars they manufacture. By doing so, they would take on the role of energy producer in addition to their role as automotive manufacturers. Thus it is possible to directly link the introduction of electric cars with the transition to renewable energy in an unbureaucratic manner – a multifunctional effect. I proposed this concept in my memorandum on greater speed for electric mobility, published in April 2010.[65]

The development towards electric mobility marks the automotive industry's entry into the field of battery production. The demand for batteries will come from the electric cars themselves, again driving energy change. The same applies to other potential storage technologies for use in vehicles or elsewhere, such as compressed air reservoirs which, in contrast to batteries, have an unlimited number of charging cycles and are entirely residue-free. Compressed air could replace batteries. Again, demand and the wide range of potential applications for compressed air will ensure its use in areas over and above the automotive sector. The vehicle industry could also differentiate its product palette within the traditional combustion engine sector, by entering into the production of mini-cogeneration units (i.e. by producing cars – themselves mini-power units on wheels) and supplementing this with the production of small stationary power units. Mass production causes costs to fall quickly, enabling the operators of these power units to undercut every conventional energy offering. In addition to mini-power units, combustion technology can also be applied in heavy commercial vehicles, e.g. for the future use of biofuels in aircraft engines.

Construction materials as sources of solar energy

Many synergies also emerge in the construction industry which is, and will remain, a major industrial sector. When entire roofs, facades and windows can be used for generating solar power and heat and providing insulation, making conventional energy costs irrelevant, then this opens up new opportunities for the construction industry. The increasing costs of traditional forms of energy, known as hidden "grey energy" in houses and which forces up the price of construction materials, makes wood, a natural product, continually more attractive as a construction material. Wood has already started to enjoy a renaissance. When timber production increases, because (as I explain in the next chapter) there is no alternative to a comprehensive reforestation programme in order to stabilize the global climate, this will only serve to increase this trend.

One role of the timber industry is to produce paper. But with the large quantities of cellulose involved, the timber processing industry will inescapably grow into the role of bioethanol producer, to cover both its own power needs and for the biofuels market, which ought to rely primarily on the processing of organic waste. The byproducts of bioethanol production can also be reused, for example as fertilizer or fuel for cogeneration plants. The more such multiple forms of use are embedded into regional resource cycles, the better they can be realized.

Diversification in the chemicals industry

The situation is similar in the chemicals industry whose most important raw material is petroleum. Fifteen per cent of global oil production is processed by the chemicals industry – 600 million tons annually. As oil becomes scarce, the price of this dirty hydrocarbon rises. It can be replaced by organic – and thus clean – hydrocarbons. It is more important that biomass is used as a raw material than burned as a source of energy, because the chemicals industry has no other alternative to this new (and renewable) raw material. In turn, this change in raw material encourages the decentralized production of chemical products. Major refineries that process petroleum derivatives can be replaced by smaller biorefineries, avoiding the need to transport biomass over long distances. Several chemical companies have already started to make this transition. Phytochemical products can not only be recycled more easily and using less energy, they can also be used as a source of energy in the form of organic waste, again starting by covering the manufacturer's own energy needs. In this way, all packing materials and plastics could become sources of renewable energy. The ideal political instrument for introducing this change would be a cut-off date, after which only organic packing materials may be used. This lightens and simplifies the work of the recycling industry, which would then only have to deal with two categories of waste: organic waste and hazardous waste. The organic waste would be used for generating energy, becoming an integral component of municipal energy supply. And instead of paying for this raw material, the municipalities who recycle this waste could even demand payment for its disposal.

Renewables-based agribusiness

The opportunities for synergies and the multifunctional use of renewables are particularly prevalent in agriculture, and are partly due to the ability to link the production of foodstuffs, energy crops and raw materials for industry. However, it should not be forgotten that bioenergy is the most complex form of renewable energy. Agriculture and forestry can lead to disastrous ecological consequences (over and above the production of energy and raw material crops) although they can also be organized in a sustainable manner. The mere harvesting of vegetation as a source of energy without replanting (e.g. by cutting down forests, typical in developing countries where, for economic reasons, the population has no other access to energy) is more ecologically damaging than fossil fuel combustion, as it leads to widespread soil degradation and the drying up of regional water sources, and thus it cannot be regarded as a renewable use of bioenergy. The same applies to agricultural production methods which rapidly exhaust the land, leading to the dramatic loss and pollution of topsoil and groundwater.

The priority of agriculture must be to produce food. Yet the waste materials from this process alone represent a huge potential source of energy which, when used to generate power, produce further waste materials which can subsequently be used as fertilizer, animal feed or for generating yet more energy. This production chain encourages the development of a recycling economy, for the more fully these raw materials are utilized, the lower the transport costs. This form of agriculture is both more ecological and more economical than its conventional counterpart. By drawing on its natural, primary source of energy, agriculture reclaims its growing, permanently sustainable – and usually subsidy-free – role as a classic "primary industry".

This presupposes that energy is generated close to the site of agricultural production, for instance by local agricultural cooperatives. But even where the production of foodstuffs takes priority, it is still possible to use productive land for several functions, for example with a second sowing or by planting energy and raw material crops after agricultural crops have been harvested. Normally these secondary crops need not be fully mature before harvesting, and nor must they be food crops. This form of crop rotation system enables agricultural production to be "multicultural". Cooperatively-run biorefineries are also suitable for recycling these agricultural products into industrial raw materials. This enables farmers to cover their own power and fertilizer requirements and take part in the first steps of marketing them, thereby reducing their own costs and increasing yields.

This shift in the organization of agricultural production requires an agricultural policy designed to meet these criteria. This policy must not shy away from limiting the production of energy and raw material crops which require excessive amounts of water, such as maize, or from making crop rotation and the multiple use of agricultural land a requirement for the production of energy and raw material crops. All agricultural waste products can either be fed back into the agricultural process or used to produce energy. Here biogas production is the "all-round solution", because every form of biomass can be subjected to an anaerobic fermentation process, with the "degassed" residue being put to further forms of use.

B. CONVERSIONS: THE TRANSFORMATION OF UNPRODUCTIVE ECONOMIC SECTORS

The concept of converting an economic sector arises from discussions of the arms industry. In the major industrial countries, this industry has been accompanied by the growth of the notorious *military-industrial complex*. The arms industry is fed almost exclusively by government contracts. It is highly technical and absorbs both a large number of specialists and a major proportion of national budgets. Many munitions are produced not primarily for reasons of security, but rather to keep the existing arms industry afloat. Governments and parliaments are being permanently pushed into new contracts with the argument that jobs need to be saved, and with increasingly complex technologies making the munitions themselves ever more expensive. Symptomatic of this was NATO's MRCA Tornado, a highly controversial fighter jet during the 1980s. The acronym MRCA – Multi-Role Combat Aircraft – was referred to by insiders as "Military Requirements Come Afterwards", indicating the extent to which its military relevance was a secondary consideration in its purchase. When it comes to economic reorientation, it is this military-industrial complex, most developed in the US, Russia, France and Great Britain, and now also in Germany, which acts as a deadweight. It systematically hinders political attempts to demilitarize, absorbs technological know-how and drives forward the technological arms race, producing insecurity rather than security. New major powers such as China and India are also seeing the emergence of their own military-industrial complexes.

Strategies for converting the arms industry are much under discussion in countries where this industry is strong. However, to achieve such a conversion the industry must focus on manufacturing products with a high technological profile and for which there is an obvious, although currently neglected, need, i.e. products that to date no industry has properly considered. As the arms industry is the focus point in maritime and aeronautic technologies, an obvious means of converting this industry would be for it to move into two areas of future significance – new forms of energy use in shipping and aviation.

Renewable energy for maritime industrial products

When it comes to discussions on alternative energy, shipping is almost a *mare incognitum,* although energy use at sea has catastrophic effects both on the climate and on marine biology. Shipping is fuelled by heavy fuel oil, the dirtiest fuel of all. It is the residue of refinery products and contains extremely high levels of sulphur. The North Sea Action Conference (AKN) stated that, "on land, ships would be hazardous waste burning plants". Shipping fleets consume on average 1 million tons of heavy fuel oil per day and emit more than 1 billion tons of CO_2 into the atmosphere. The effects on marine biology of leakages or tank-cleaning on the high seas stretch right into the food chain, are unquantifiable and cannot be "economically" compensated for in any manner. Therefore it is high time for a political programme for the development of emissions-free ships that run on renewable energy. In order to simultaneously make this a method of converting the arms industry, companies involved in constructing warships should be commissioned to do this. Initial financing for this development must be raised by reallocating portions of the defence budget.

The potential for running ships on renewable energy extends right through to energy-autonomous ships and includes very large container and passenger ships. An example is the "Skysail", a large foil kite attached to a cargo ship, where wind power can replace the ship's engine on a ship of 120,000 gross register tonnage. Another option is the direct use of wind energy onboard the ship, as well as the production of solar power on the vessel's sides.

This power could be used to produce hydrogen by electrolysis, with the hydrogen being used to drive the ship's engines. Wind power plants near harbours, used to generate hydrogen by electrolysis, can help by reducing the energy loss inherent in transporting hydrogen. Practical examples are also given by submarines which, for many years, have also been driven by fuel cells. Passenger ships carrying a large number of passengers could convert organic waste using biogas plants, generating power to drive the ship and meet its energy needs. A ship as an autonomous power plant! Other ideas include drive technologies which run on vegetable oils – when ships such as these spring a leak, no longer is the ocean contaminated, instead fish and other marine inhabitants are provided with additional nutrition!

What applies to the maritime shipping industry (as well as ideas for changes to warship construction) also applies to the construction of boats and ships destined for inland waters. It has long been possible to drive passenger and pleasure boats using electric motors, the power being produced by solar cells installed on deck (as demonstrated by the pleasure boat *Solon* on Berlin's Spree canal). Likewise, it has also long been possible to drive ships and motorboats with engines run on vegetable oils, for purposes of climate and water protection. These changes do not require public research and development programmes, simply the political courage to stop authorizing new craft driven with fossil fuels.

Thus shipping returns to its original sources of energy which, for thousands of years, were solar: from ancient sailing ships to modern solar shipping.

Energy change and the structural transformation of aviation

The second extremely neglected approach is that of energy change in aviation. In the emissions hierarchy, aviation fuel – kerosene – ranks above heavy oils and below diesel and petrol. However, aviation emissions, particularly at altitudes of over 8,000m, have a significantly more damaging effect on the climate than those from vehicles on Earth: they degrade only slowly, lead to the build up of ice crystals in the atmosphere and prevent both solar radiation reaching the Earth and heat radiating back from the Earth. Powering aircraft using renewable energy is an enormous technological challenge, one which will lead to a fundamental transformation of the aviation industry. Biofuels are the most likely and obvious alternatives to fossil fuels for aviation. But development shouldn't stop here, especially considering the growth in international freight and charter flight traffic.

Therefore it is incomprehensible that the aviation and space industry, as well as governments, pay almost no attention to aviation technologies. Despite there being a Social Democrat/Green coalition at the time, in Germany a promising opportunity was missed in the form of the Cargolifter airship, a private initiative which, despite unsure yields, managed to secure financing via shareholder funds raised on the stock market. The concept behind the Cargolifter is to load freight at its point of production onto an airship parked in the air immediately above, subsequently unloading it directly at its final destination. Despite the slow speed of airships, it is still quicker to transport freight by air than over land or by rail – the longer flight times are more than compensated by avoiding the earthbound transport routes to and from traditional freight aircraft. The world's largest production hall, south of Berlin in Brand, in the state of Brandenburg, had already been erected and the project subsidized by Brandenburg's state government. However, this subsidy, together with around €300 million raised on the stock market, was still insufficient. Compared with the billions spent in developing new, conventional aircraft models, this sum was clearly too low.

Not only do airships use extremely little energy, allowing them to be driven using electric motors, their decisive advantage is their almost complete independence from physical infrastructure. As well as transporting freight cargo, they are an attractive tourism alternative, particularly for charter traffic. They deserve significant political attention and support, if only as a conversion project for the aviation industry.

Coal products for industrial needs

A conversion strategy is also needed for the coal industry. It is ridiculous to wish to use up all available coal deposits in order to generate electricity; we can live without coal for generating power and heat, but it remains vital to the steel industry and for its many applications in the form of carbon fibre materials, rendering the term "free carbon economy", commonly used in debates on the global climate, a misnomer. Coal, in the form of coke, is needed to produce pig iron. Coal is heated in coke burners under airtight conditions to over 1,000°C. During this process the volatile components leak out: hydrogen, methane, ammonia and others – the so-called "white side" of the coking process. These gaseous substances were formerly used as coke oven gas or town gas.

The remaining coal reserves must be used for producing steel which will continue to play an important role: compared to concrete, steel has proven to be a more stable and flexible construction material, one which is less energy intensive and, above all, recyclable. Steel bridges are more durable than concrete bridges and their lifecycle costs are lower. As the coke won from coal is needed to produce steel, closing Germany's most modern large coke burner in Dortmund in 2000, before dismantling it and shipping it to China in 2003, was an act of folly.

Materials made from carbon fibre have a multitude of applications and this constitutes the second area for change in the coal industry. Compound materials made from carbon fibre are as stable as metals but significantly lighter and require less energy to produce. They are increasingly used in the production of ships and automobiles and play an important contribution to reducing energy needs. The coal industry must be focused on supplying the steel industry and producing carbon fibre materials, and this may perhaps help reduce the industry's resistance to energy change. Seen from this perspective, together the coal and steel industries can play a role in the transition to renewable energy, for it is also possible to use hydrogen, produced using renewable energy, in the steel manufacturing process.

However, a precondition for supplying hydrogen is to avoid the need for expensive and loss-making infrastructure and transport. Accordingly, hydrogen should be produced in harbours, so that ships can be refuelled directly. This hydrogen could also be used to store part of the power generated from renewable energy. This development begins with power generated from renewables, or from coke oven gas used in cogeneration plants and reserve power plants for renewable energy respectively. One element which should not be ignored here is the colliery gas in decommissioned mines which can be used to generate power, and thus is included in the Renewable Energy Sources Act (EEG) although not itself a renewable energy.

Converting the power companies

Conversion is also conceivable for the conventional power industry. This industry is faced with the decision of whether to artificially prolong the conventional system of energy supply, or to play a role in shaping the clearly recognizable path towards highly flexible, decentralized and rationalized energy supply. The latter implies freely relinquishing its monopoly position and, of course, revenue losses. In return, it will not be completely forced out of the market, nor will it suffer a fundamental loss of acceptance. The compelling reasons for rapid energy change can't be falling on deaf ears throughout the entire conventional energy industry.

The industry could extract itself from its structurally conservative and egotistical business position by promoting decentralization, with power companies transforming themselves into holding structures for companies which operate independently at local and regional level. However, we currently see no example of this in action. The other option is gradual withdrawal from the power industry itself, and entry into other fields of business activity. Here we have the example of the former German company PREUSSAG, once a national mining, coal and steel industry, which now operates under the name TUI in the tourism industry.

C. LIBERATION: THE OPPORTUNITY FOR DEVELOPING COUNTRIES AND A "DESERT ECONOMY"

The disastrous situation in which developing countries find themselves cannot be understood without reference to the conventional energy supply system, for they are its first major victim. Their energy tragedy began from the point at which a centralized energy supply structure, one which had evolved gradually in industrialized countries, was implanted in developing countries. This energy-sociological "fall from innocence", whose ramifications continue to the present day, has led to larger cities being supplied with power generated by large-scale power stations – but not the rural areas where often more than 90 per cent of the population live or lived because the economic power of these countries was insufficient to support the construction and operation of overland networks. With major cities modelled on those of industrialized countries on the one hand, and traditional agrarian structures on the other, this has resulted in increasing social division and the rapid growth – historically unique – of cities with mushrooming slums.

Developing countries pay the same price for imported energy as rich industrialized countries, but with gross national product (GNP) per head at around 10–20 times lower, they have become caught in a vicious circle. They are forced to import modern power plant technologies although these plants do not supply the majority of the population with the power they generate. Usually they must import the energy to drive these plants, but without having the purchasing power to do so. In many developing countries, energy import costs are greater than their total foreign exchange revenues; and even so, the majority of the population are excluded from energy supply. Every additional form of economic activity needed to escape from this situation requires the use of more energy and thus more imports, so that potential yields are already devoured by the additional energy costs. When, in 2007, the price of oil shot up, the developing countries' bill for importing oil rose by around US$100 billion within a single year, i.e. more than the industrialized countries' entire annual development aid budget (around US$70 billion). Despite this, energy economists continue to persuade the governments of these countries that the transition from their long unaffordable imported energy to local renewable energy represents an economic burden.

At the same time, dependency on conventional energy supplies prevents developing countries from striking out in the direction of the most obvious development: from their role as suppliers of agricultural produce and minerals to processors of raw materials and exporters of semi-finished and finished goods. In order to do so, agriculture must be more productive and carried out in a socially acceptable manner, i.e. not through large-scale agribusiness, but by small farmers organized into cooperatives. However, this requires that power be rapidly made available in rural areas. There must also be enough energy for processing the raw materials – energy which doesn't have to be expensively imported. In short: these countries must mobilize their own reserves of renewable energy.

A single example illustrates the energy trap in which developing countries find themselves: 50 per cent of global bauxite reserves (the most important raw material for producing aluminium) are located in Guinea, in Africa. However, because of a lack of energy, they can process only 2 per cent of the bauxite they extract annually. Thus developing countries find themselves in a notable dilemma: on the one hand, the conventional energy system has left the field free for developing a renewables-based system of power supply; on the other, many of their decision-making elites continue to strive for a centralized power supply system although, because of increasing fossil fuel prices, this is increasingly impossible to achieve.

Even if most developing countries need to import the technologies for using renewable energy, this would still offer them two decisive advantages: they could increasingly avoid the need to import energy and would boost their own economic standing by processing their own raw materials. Thus they would be able to begin their own investments into renewable energy and replace energy imports without being dependent upon development aid. They could expand these activities by using their national banks to provide loans for investments designed to replace energy imports. They would be financing independent initiatives using the foreign exchange revenue no longer needed for energy imports. They also have the power to direct development aid from third parties towards renewable energy, and to use cooperative banks to make microcredit widely available along the lines of the Grameen Shakti Bank in Bangladesh.

"Desert Economy"

The alternative to the DESERTEC project (or similar attempts to use the potential offered by solar radiation in desert regions for international energy supply) is for desert countries to use this potential for their own benefit. Solar and wind power allows desert states to meet their growing energy demands and drive their economies forward. Having direct access to a source of power is of far greater economic value to these countries than potential future power exports to consumers in far-off lands (a perspective which, as shown in Chapter 3, is riddled with uncertainty).

For desert countries with petroleum and gas reserves such as Iraq, Kuwait, Saudi Arabia, Abu Dhabi, Libya, Qatar or Algeria, using renewables to produce power to cover domestic demand is still economically relatively uninteresting. In contrast, for Morocco, Tunisia, Egypt or the countries of the Sahel, using renewable energy has become a matter of economic survival. The countries of the Sahara have many mineral resources which, to date, they hardly process: sand and kaolin for the production of silicon, glass and ceramics; barium for the production of vacuum tubes, spark plugs and plastics; lead for batteries or radiation protection materials; calcium for alloys, cement or as a reducing agent for rare earth (vanadium, thorium, zirconium, yttrium); fluorine used in aluminium production, electrolysis systems, surface treatments, the production of paint, enamel and toothpaste; cobalt for magnets, steel alloys and for medical radiology, or as a pigmentation; magnesium as a light metal and for medical technologies; manganese for steel alloys and titanium for shipping and medical technology; phosphates for fertilizers and detergents, etc. These mineral resources are mainly located in the mountainous areas of the Sahara, such as the Atlas and its foothills, the Hoggar (southern Algeria), Tibesti and Ennedi (Chad) or in desert sands. Most are excavated by international mining companies before being transported to Europe, North America or Asia for processing. Due to a lack of energy, as well as water, most desert countries are unable to process these resources themselves. However, with the help of solar and wind power, both these deficiencies can be overcome. Desalinization becomes possible, simultaneously providing the population with drinking water and water for intensifying agricultural activity.

Generating electricity using renewable energy provides desert countries with comparative cost advantages. It makes it easier for them to increasingly process their own mineral resources rather than exporting them to industrialized countries – their classic role, one imposed upon them during colonial times and practised to the present day. The economic future of these desert countries is bound up in this energy supply. An example: Morocco is North Africa's most stable country and is relatively well-developed, with a rapidly growing population of 34 million. But it currently imports 96 per cent of its energy requirements – oil, gas and coal – for which it must use more than a third of its annual foreign exchange revenues. If the EU supports Morocco and the desert countries in making the transition to renewable energy, then this will contribute greatly to developing these countries and making them more stable. This measure alone would prevent growth in the numbers of economic refugees fleeing from these countries.

D. PREVENTION: FUTURE OPPORTUNITIES FOR ENERGY-EXPORTING COUNTRIES

For a long time it was the Organization of the Petroleum Exporting Countries (OPEC) that blocked all international efforts to break up the conventional energy system. This was evident at the Rio conference of 1992 where they rejected the suggestion that Agenda 21 should highlight the particular importance of renewable energy for global climate protection and sustainable economic development. The OPEC countries feared that a focus on renewable energy in industrialized countries would damage their own export interests and went as far as to demand that industrial countries that made the transition to renewable energy should pay financial compensation to the petroleum-exporting countries. At the world climate conferences which began in the mid-1990s, one could see how lawyers, paid by the transnational petroleum companies and OPEC countries, continuously supplied government representatives with catchwords during the conference, aimed at preventing agreement on climate protection being reached.

Some of these countries have only recently understood the need to prepare for the "time after", when their energy reserves have become exhausted. There is an attempt to follow the example of Norway which, for many years, has put aside a large proportion of its income from the export of North Sea oil and gas as a form of future fund. In most energy-exporting countries the entire economy is based on such export revenues. This is the case not only for the Arab oil and gas exporters, but also for countries such as Russia, Venezuela, Mexico and also Australia, which exports coal. If these countries fail to restructure their economies in good time, then they all run the risk of economic collapse once their reserves have been exhausted.

However, the interests of the energy-exporting countries are extremely ambivalent. Their current economic interests motivate them to delay energy change for as long as possible – a goal they share with the energy companies.

In several of these countries, a formal distinction between government and the energy companies is lacking, with petroleum and gas extraction having been nationalized. Yet it is these countries that have greater financial room to manoeuvre when it comes to investing in renewable energy. Influenced by concepts such as DESERTEC, several are considering continuing their role as energy exporters, but in the form of power generated by renewables, and thus retaining a centralized energy supply structure. However, as we have seen, this concept and its likelihood of ever being realized are extremely questionable, and runs counter to the specific characteristics of renewable energy and its related technologies.

If, and how, the current fossil fuel-exporting countries will take part in the transition to renewable energy is a central question because, in light of their current revenue situation, these countries are unlikely to be satisfied by focusing solely on meeting their domestic energy needs. Their important role in the global economy, one achieved as the result of their energy exports, is long since reflected in their systemic investments. The energy-exporting countries have used their export revenues to buy heavily into businesses in the importing countries. Countries such as the United Arab Emirates – whose strongest pillar is the emirate of Abu Dhabi where the majority of its energy resources lie – and also Bahrain, Qatar and Saudi Arabia, have discovered their future interest in renewable energies, as has the Caucasian petroleum and gas exporting country of Azerbaijan, which is not a member of the OPEC. The World Future Energy Summit, hosted by Abu Dhabi since 2008, is a signal of the energy exporting countries' recent focus on renewables.

It is one of the ironies of the 20th century that here a further fundamental change in the roles of the energy exporting and importing countries is looming. The majority of the countries listed above were formerly colonies, their colonial powers using them as a resource base. After independence, economic dependence on the international power companies and former colonists initially remained. However, now the roles have been swapped. With the extracting countries' growing confidence, their joint solidarity in the form of the OPEC and, above all, the growing energy needs of the importing countries, it is the importing countries themselves who have become increasingly dependent on their former colonies. Now it is the exporting countries (those who have saved their revenues) that are the few financially solvent states, and it is the governments of the former colonial powers who make regular courtesy visits to ask for handouts or to secure contracts for their domestic companies. If the energy exporting countries wish to preserve their newly-won dominance in the global economy, then they must grab this opportunity now. However, this assumes that they will use the huge range of investment opportunities that will arise during the dying phase of the conventional energy system to focus their future economic strategies on the production of renewable energy technologies – without waiting for all available conventional energy resources to first be exhausted.

6

AGENDA 21 RELOADED

Global federal initiatives for energy change

Having experienced the repeated failures of world climate conferences, we should now draw a line under all previous attempts at global climate protection and dare to make a radical new start. At world climate conferences two complex approaches – minimum obligations, and CO_2 offsetting and trade – have dominated global energy discussions. However, focusing solely on the problem of CO_2 is too one-sided and, above all, fails to properly take into account the key question of energy supply. This restriction, as well as the shifting of decisions on energy supply (vital to every national economy) to the global negotiating level, has simultaneously led us to lose sight of the important stimuli which arose from the promising 1992 Earth Summit (UNCED) in Rio de Janeiro, with its Agenda 21.

Agenda 21 listed all the problems with which, sooner or later, almost everyone will be faced, highlighting the need to act according to new principles. These are either directly, or indirectly, connected with energy supply: from climate change through to destruction of the ozone layer, from encroaching deserts through to soil erosion, from forest decay through to water pollution, from the dangers to health which result from this environmental degradation through to the loss of species diversity, from biotechnological risks through to the burden of waste, from poisonous chemicals through to the destruction of marine life. It also lists the consequences for human living conditions, which are also inevitably linked to forms of energy supply: growing poverty, the loss of food security, environmental quality and living standards. The Rio conference was the first and most spectacular Earth Summit, both for governments and for the NGOs at their Global Forum. It also passed a framework convention on climate protection and formulated fundamental principles for new political and economic action.

However, attempts to derive tangible and contractually-binding obligations and start joint action programmes have been unsuccessful. But what was carved in stone at this Earth Summit was Principle 1 of the Rio Declaration: "Human beings are at the centre of concern for sustainable development. They are entitled to a healthy and productive life in harmony with nature." This *sustainable development* must "be realized so that the developmental and environmental needs of current and future generations are met in a just manner". Sustainability became the guiding principle behind ethical responsibility, motivating many people, unprepared to wait for new laws or international agreements, to undertake their own initiatives. "Think global, act local" became their battle cry. The many sources of local activity should develop into a widespread movement which finally involves everyone.[66]

This participatory and emancipatory approach is exemplary, turning the "spirit of Rio" into practical reality. The majority of successful environmental projects are based on initiatives such as these. The opposite tactic is global political centralism which has proven unsuccessful over the years. It leads to a series of world conferences at government level which produce big words but little action, and reveal, more than anything else, the extent to which governments are embedded within the power structures of retarding interests. Over the years, the guiding principle of sustainability has been systematically diluted and its content compromised; similar political approaches have been mired in bureaucracy. Against their better judgement, many NGOs allowed themselves to be integrated into this process. Sustainability was reduced to a subordinate role, and forced to make way for the free market agenda. The least sustainable elements of world civilization are nuclear power and fossil energy. The most important prerequisite for achieving sustainability is the transition to renewable energy.

Now, after almost two decades of failure (by no means a coincidence), it is high time to reanimate Agenda 21 and rediscover forms of action which are socially inclusive and thus promising. This requires as many individual initiatives as possible, and no more centralized initiatives than absolutely necessary. The same applies to international policy – as many national initiatives as possible and global initiatives only where these cannot be supported by individual states alone. It means political competition to drive forward energy change, which is incompatible with internationally standardized arrangements. All that can be "standardized" is the physical goal of shifting mankind's energy supply to renewable energy – but not the competencies and precise starting conditions for achieving this. A political world order can be based only on a federal structure.

Although at first sight, the core concept that global problems can be overcome only by means of a standardized, politically consensual approach might appear logical, it is, however, politically naive as well as deeply undemocratic. The disastrous consequences of standardized economic planning are demonstrated by the command economies of the former Eastern Bloc and by the current devastating effects of neo-liberal dogma as it attempts to standardize the global economy. A "federalist world order", as described by Otfried Höffe in his book *Democracy in an Age of Globalisation*, is not only more democratic and humane, but also closer to each of the tangible problems which need to be overcome.[67] It is more flexible and open to new ideas which, when successful, can be used as models for others. No one should wait with their own initiatives until others are prepared to join in – when dealing with burning issues, it is even irresponsible to do so. Thus deciding between unilateral and multilateral concepts is irrelevant. The world conferences which followed the Rio Earth Summit were fixated on multilateral and consensual ideas and underestimated, or even denied, the effects of unilateral efforts. In contrast, successful unilateral concepts can stimulate international wave-like movements which have a greater effect than protracted negotiations on harmonized multilateral policy.

The so-called Copenhagen Accord, the final declaration of the world climate conference of 2009, created the approach of "pledge and review", submitted by President Obama, and was correct in principle. It requires individual countries to set their own quantitative goals for reducing CO_2 emissions, and forces them to justify themselves internationally and morally where they fail to achieve them, shifting responsibility away from international agreements and back to individual states. This principle should be seen as the starting point for a long overdue revision of previous world climate policy and adapted productively. President Obama's suggestion also demonstrates, in an exemplary manner, the actual room permitted him to manoeuvre by the US Senate; he was unable to promise more, as experience shows that he would be unable to count on agreement for a more binding contractual obligation. The same is demonstrated at all world climate conferences: the more centralized the proposed solution, the more disruptive the national power structures of conventional energy supply become, and this leads to a dilution of goals right from the outset. Thus energy change in the US will not be the result of federal initiatives, but rather those of a growing number of cities and individual states.

The multilateral effects of unilateral initiatives

Our first task in introducing energy change is to eliminate structurally conservative influences, for the more that decisions are determined by consensus, the greater is their power to obstruct, enabling them to effectively determine both the speed and goals of energy change. However, the more these influences are counteracted by autonomous, unilateral initiatives, the greater the extent to which the speed of change is determined by avant-garde forces who set their own standards. This has been demonstrated by the unilateral implementation of the German Renewable Energy Sources Act (EEG): it arose independently of world climate conferences, was never the subject of negotiations and, together with similar laws introduced in many countries, has effectively lead to a greater reduction in CO_2 emissions than the Kyoto Protocol which was initiated at world climate conferences and focused on emissions trading. Elinor Ostrom describes the EEG as an outstanding example of an international design competition with a recognized goal. She regards a fixation on global effects as the wrong approach, even for climate policy. The "global governance" approach to which many have sworn allegiance, and which is intended as a substitute for a non-existent world government, has largely failed: the prospect, as described by Thomas Fischermann and Petra Pinzler in their article on the illusion of a unified world, published in the German weekly broadsheet *Die Zeit*, of "governments and national administrations becoming ever more coordinated, and supported by global concerns, NGOs, a cosmopolitan host of academics and other experts. In the end, the world would be governed by a mixture of negotiated agreements, contracts, the proper institutions and many conferences".[68] However, the true outcome has been political paralysis.

Only since the frustrating results of Copenhagen, and the equally meagre and ambivalent results of the truly inflexible "flexible instrument" of the international CO_2 certificates market, have those who for too long have accepted the multilateral concept of obligations and implementation behind world climate conferences been forced to reconsider. In the spring of 2010, Connie Hedegaard, the European Commissioner for Climate Action, officially admitted that the EU system of emissions trading had failed to achieve its intended purpose and had not stimulated climate protection investments.[69] But the EU Commissioner did not dare recommend abandoning the concept of emissions trading, one with the established status of sacred cow.

The German Advisory Council on Global Change (WBGU), under the chairmanship of Hans Joachim Schellnhuber, director of the Potsdam Institute for Climate Impact Research (PIK), has also undergone a similar change of heart. In its "policy paper" on "climate policy after Copenhagen", the Council criticizes the "principle of consensus as hampering decision-making" and recommends "European-wide feed-in tariffs for renewable energy" with the aim of achieving "full European supply with renewable energy by the year 2050", the "implementation of a high-tech strategy", "initiatives by cities and municipalities working together in climate alliances, as well as input from businesses and players in civil society" and "sub-global alliances of climate pioneers" with "privileged partnerships", which can function as "the self-reliant motor for a new multilateralism in climate policy" in order to stimulate a "competition in green innovation", by means of "polycentric" initiatives. Notwithstanding, not only does the Council remain loyal to the concept of cap and trade with agreed upper limits, it also recommends the expansion of the system – although it admits that this will serve to "further increase the complexity of existing emissions trading". It even recommends that cap and trade be expanded to cover "all businesses", going so far as to "establish monitoring and reporting systems for measuring emissions in wooded areas".[70] Thus the WBGU closes its eyes to the necessity of slaughtering the sacred cow, even accepting conceptual contradictions in order to do so.

In terms of content, the undignified bargaining over set quotas is absurd – especially considering Nicholas Stern's widely accepted calculations that the economic costs of ignoring climate protection are significantly higher than those of avoiding CO_2 emissions. Measured in terms of the political effort required to establish a system of emissions trading, its lack of success, long-windedness and minimal impact (even should it eventually succeed) are obvious. International emissions trading has become an end in itself. However, if governments are determined to retain such a system, then they can do this within their own economic region, where it can be implemented in a slightly more transparent manner, one safe from misuse. If nothing else were to happen, international emissions trading would represent only a minimal improvement over the status quo. But "do nothing" behaviour is no longer an option. We need emphatically more activities to promote global energy change, based on a catalogue of tasks extending beyond climate protection – a form of "Agenda 21 reloaded", one that draws a line under the concepts which have diverted us from this goal since 1992.

A world conference for sustainable development and climate protection

CO_2 avoidance is not always synonymous with sustainable development and the transition to renewable energy, as shown by the concepts favoured at world climate conferences. However, every single investment in renewable energy is synonymous with CO_2 avoidance (with the exception of bioenergy, where this only applies under particular conditions). Yet a sustainable economy is unthinkable without a change to renewable energy. Consequently, we must transform world climate conferences into *world conferences on sustainable energy supply and climate protection.* This would be tantamount to a political and doctrinal focus on renewable energy and the demands of technical and natural climate protection. Ideally, this would take the form of an *annual UN Special Session.* As such, it would no longer be held back by laborious and drawn-out international treaty negotiations. Instead, as a representative global forum, it could finally debate the paths to sustainable energy supply within a suitable framework – with sustainability its principal political goal. At the same time, it could finally instruct specialized UN agencies to pay more attention to the goals of sustainable energy supply, something that previously only the United Nations Environment Programme (UNEP), has done.

We can do without "target and time" resolutions, i.e. obligatory quotas for reducing emissions and introducing renewable energy within set time limits. Once we start dealing with international obligations that impact on the overall economic structures of each nation, endangering its influential and structurally conservative powers, then the paralysing haggling for consensus inevitably continues. Policies that encourage independent energy change initiatives are far more important than quantitative targets. What we lose by abandoning obligations (where we can achieve the minimum at most), we win in the form of a focus on the real problems and motivation for new activity. We could retain the most important structures of world climate conferences, namely the UN

Climate Secretariat with the Intergovernmental Panel on Climate Change (IPCC), whose reports have created and maintained global awareness of the dangers of climate change. More important than painstakingly negotiated quotas are unvarnished analyses of the problems which help to clarify goals; the free exchange of experience and the cultivation of ethical standards; the establishment of international framework conditions that simplify energy change for everyone and overcome the obstacles lurking in scores of international treaties; the obligation on global government organizations to pursue the goal of sustainable energy supply; the creation of unbureaucratic and widespread initiatives for investment. The leitmotifs are: greater direct, horizontal international cooperation in place of one global regime for all; aids to self-help and individual responsibility; priority for rapid action rather than protracted negotiations, in our own self-interest and for the common good.

A. 350PPM: RECAPTURING CO_2 BY EXPANDING AGRICULTURE AND FORESTRY

The much discussed "two degree limit", which marks the barely tolerable upper limit to global warming and accepts an increase of CO_2 in the atmosphere from its current level of 385ppm to 450ppm, is a partial capitulation. Our goal must be to return to the climatically stable value of 350ppm, the level prior to the start of our unsustainable fossil energy age. In 2009, the US environmental activist Bill McKibben started a "350ppm" initiative, calling this "the planet's most important number", as "an even safer upper limit for CO_2 in the atmosphere". The initiative is based on the research of the NASA climate scientist James E. Hansen who concludes that "when mankind wishes to preserve the planet in a state similar to that in which our civilization developed and to which the entire life of our planet is optimally adapted, then we must reduce the CO_2 content of the atmosphere from its current level of 385ppm to 350ppm".[71]

Converting world energy supply to 100 per cent renewable energy will not be sufficient to achieve this 350ppm target – it can only prevent CO_2 levels in the atmosphere from increasing further. To bring these levels back down to 350ppm we need to mobilize the powers of nature, above all by expanding forested areas and upgrading soils to use them for carbon storage. At world climate conferences to date this has only been attempted by means of the REDD concept (Reducing Emissions from Deforestation and Degradation) which is limited to preventing the further loss of forestry and other vegetation. Slash-and-burn leads to the loss of 130,000km^2 of forest each year and accounts for around 20 per cent of annual CO_2 emissions. The WWF calculates that it would cost around US$20 billion to US$33 billion per year merely to halve these losses.[72] Even so, this would be the most cost-efficient means of reducing CO_2 emissions. Yet the question of financing remains. Suggestions range from including reforestation or the reduction of slash-and-burn into emissions trading systems, through to a fund provided by the industrialized countries. In Copenhagen, promises were made to the tune of US$30 billion for the period covering 2010 to 2012.

However, efforts to prevent increased CO_2 emissions cannot be limited to incentives for protecting vegetation. The excess levels of CO_2 in the atmosphere must be reduced by means of global CO_2 recapture initiatives. But the REDD is also absurd for other reasons: why should governments be paid for preventing degradation to their own areas of vegetation – especially the tropical rainforests – when it is in their own interest to maintain this vegetation and use it in an economically sustainable manner? Couldn't this lead to official announcements of further planned deforestation, simply to collect payments for subsequently agreeing not to go ahead? And how realistic is it to expect that such an expensive and possibly long-term programme can be funded, bearing in mind the efforts needed to raise the initial funds in Copenhagen? Assuming that rejecting destructive, and adopting sustainable, structures generates new opportunities for economic growth, and bearing in mind the "spirit of Rio" which encourages cooperative and participatory initiatives, the more sustainable option is clearly a major global reforestation initiative.

Apart from the desert regions, geographically the opportunities for reforestation are almost unlimited. Even in desert countries, this is at least possible in coastal areas when carried out in conjunction with desalinization and irrigation systems. Reforestation projects such as this bind the soil and revitalize the natural water balance, and by reintroducing agricultural production methods they are also of social value. In contrast to such programmes, and to those which involve the planting of short-term energy crops, projects to capture CO_2 by means of reforestation serve to bind CO_2 over the long term. A hectare of new woodland can absorb 300 tons of CO_2 per year for a one-off cost of around €400 per hectare. After only a few years, it will generate annual forestry yields sufficient to offset the original planting costs.

There are several ways of binding CO_2 in soil. One is the hydrothermal carbonization process for the rapid production of humus, developed by Markus Antonietti, head of the Max Planck Institute for Colloids and Interfaces in Potsdam.[73] This is a high-speed method of creating carbon, one which nature requires hundreds of thousands of years to complete. Organic material is transformed into carbon under conditions of high pressure and heat (as in a pressure cooker) in specially designed carbonization plants. The carbon is then added to the soil as a fertilizer where it generates humus. In this way, CO_2 is bound to the soil. Adding an average of 20 tons of carbon per hectare of agricultural land serves to double or even triple soil yields. These carbonization plants currently cost around €100,000 – similar to a piece of large agricultural equipment – but once mass produced, the cost will automatically fall. Compared to the increased yields and fertilizer savings, these plants are highly productive and simultaneously generate useful heat. A further method of binding CO_2 in soil by means of humus production is the "terra preta" process – the production of humus from charcoal, organic waste and excrement. Both these approaches can only be realized in a regional context. Hans-Josef Fell has calculated that, by combining a global reforestation programme with the Antonietti process of hydrothermal carbonization (using 8 million plants worldwide), global CO_2 levels could be reduced to 330ppm by 2030 if we use exclusively renewable energy from now on.[74] This last condition is unachievable, even with rapid energy change. Yet the complete transition to renewable energy is possible within a quarter of a century on a global scale, making the target mark of 350ppm an achievable goal.

Both reforestation and large-scale humus production, neither reliant on monoculture, protect the climate and stabilize the natural water balance. They can be applied everywhere and rely on the involvement of many activists. Even without subsidies, they open up new market opportunities for agriculture and forestry. In particular, they promote the use of wood as a construction material, one that binds CO_2 over the long term and can largely replace its rival cement (representing a further reduction in CO_2 emissions). Yet both initiatives require an initial political push: agricultural initiatives need training and related loan programmes, reforestation initiatives the allocation of planting areas, with regulations stipulating the planting of native tree species and mixed forests.

The success of a tree-planting project does not depend on funding, as demonstrated by the historic example of the Civilian Conservation Corps (CCC), set up by President Franklin D. Roosevelt in the 1930s. George C. Marshall, responsible for the post-war European Recovery Program, or Marshall Plan, which was financed by the US, was charged with running the project. Between 1936 and 1941, the CCC mobilized hordes of young volunteers (a total of around 800,000 over the lifetime of the project) to undertake voluntary reforestation work. In just a few years, the forested areas of the US expanded dramatically. President Kennedy also used the CCC as a model for the Peace Corps he set up in 1961. There are current examples of social reforestation projects in China, where local mayors and government members work on reforestation projects together with large numbers of citizens at weekends. Such projects can be carried out equally well by communities and schools. A good example is the Tree for Tree project, established by school children and supported by the UNEP, whose target is to plant 1 million trees in every country of the world. Their motto is: "Talking alone will not stop the glaciers melting: Stop talking. Start planting."[75]

Reforestation can also be a productive task for armies during peace time, not only in developing countries, and especially for larger reforestation projects carried out away from residential areas. An early example is the work of French Emperor Napoleon's Engineering Corps, engaged not in reforestation projects but in constructing shipping canals and irrigation systems.

Sustainable political frameworks

When it comes to natural climate protection measures, a political framework is essential if sustainability is to be guaranteed. If our use of bioenergy is not synchronized with replanting of the same order of magnitude and CO_2 balance, then we can hardly speak of sustainability. The same applies to cultivation methods which result in the erosion of soil humus and damage to groundwater. The physical law of sustainability must rank higher than any market regulation and take precedence over international agreements. It must be embodied as a guiding principle within each nation's legal framework.

The primacy of the principle of sustainability can also be justified within the framework of world trade agreements. This demands adherence to the fundamental principle of mutuality, i.e. sustainability regulations must be applied not only to imports, but also to national products. Thus the UNEP and the UN's Food and Agriculture Organization (FAO), in the form of a UN Special Agency for Agriculture and Food, should be tasked with defining and drawing up a Code of Conduct for the sustainable use of biomass. Although not itself an international agreement, such a recommended course of action makes it faster to achieve and could serve as the basis for drawing up national regulations. Once a codex such as this is adopted by major countries or the EU, its impact automatically becomes international.

B. "ZERO INTEREST" FOR ZERO EMISSIONS: DEVELOPMENT FINANCE FOR RENEWABLE ENERGY

Perhaps the most effective means of stimulating international investment in renewable energy would be for public financial institutions to issue "zero-interest loans". This would enable us to remove two investment hurdles: the economically and ethically intolerable price advantage awarded to conventional energy (whose social costs are paid not by the providers but imposed upon society), and investor reluctance when it comes to the initial costs (although investors usually fail to include the permanent avoidance of fuel costs in their calculations). In this sense, offering zero interest for zero emissions is not a subsidy, but rather a premium for avoiding social harm. Thus zero-interest loans could be provided either by national banks, who are compelled to do so by political decree, or in the form of normal bank loans, with the state financing the difference between the zero and the standard interest rate.

The potential effect of such an instrument (whether in the form of an actual zero-interest loan, as a minimal handling fee, or a low-interest loan) is demonstrated by the low-interest programme offered by the post-WWII Marshall Plan for European countries for which credit portfolios worth billions were made available. In Germany, the potential success of such programmes is indicated by the credit programme run by the national Reconstruction Credit Institute (KfW) which, with its future infrastructure programme, stimulated a wave of investments in environmental protection at the start of the 1980s. This laid the foundations for an environmental technology industry, one which has risen to international prominence and generated more than 1 million new jobs. The 1999 German 100,000-roof programme for solar power units – the world's first mass programme aimed at promoting photovoltaic technology – began with a zero-interest loan with a so-called mezzanine component, requiring the first repayments only in the third year. These types of loan characteristically have longer credit periods.

Zero-interest loans would effectively wipe out the credit industry, at least temporarily. Only the actual loan would be paid back, making investments in renewable energy significantly easier because the loan can be paid back using the savings made on fuel costs. Without impinging upon the public purse, such loans could be made available directly from central banks – in Germany the Bundesbank, and in the eurozone countries, the European Central Bank (ECB). Using a similar method, during 2008/2009, the ECB lent money to suffering banks at interest rates of between 0.5 per cent and 1.25 per cent, for the "systemic" reason of bank restructuring. It was an exceptional response to an exceptional challenge. Accelerating the pace of energy change is also the response to a dramatic situation, and the systemic reasons for doing so are no less substantial. One sentence, uttered on the occasion of the banks being saved, sums it up nicely: "If the climate were a bank, it would have already been saved."

Starting an international initiative enabling developing countries to invest in renewable energy by means of zero-interest loans would be a political task for international development banks – for the World Bank, the European Investment Bank (EIB), the European Bank for Recovery and Development (EBRD), as well as for the African, Asian and American development banks. The initiative could be backed by the International Monetary Fund (IMF) where larger credit portfolios are needed, and these could be supplemented by national development aid budgets. This would not require an international agreement – a joint pact, agreed at a G8 or G20 Summit, or resolutions passed by the governing bodies of banks, would be sufficient. Financial irregularities, such as those associated with emissions trading, could be avoided because these loans are based on concrete calculations and used for new investments. In developing countries, which primarily need microcredit for rural areas and potentially affect 2 billion people, international development banks will need to cooperate with local credit organizations. This opens up many opportunities for becoming involved.

C. HUMAN POTENTIAL: INTERNATIONAL EDUCATIONAL INITIATIVES AND THE ROLE OF IRENA

It is mankind itself which offers the most important and greatest potential for energy change. Although there is a widespread social movement for the transition to renewable energy, it can only be implemented with the aid of specialist competence. Even if, overnight, all resistance to renewable energy fell away, clear political priorities were pushed forward and sufficient investment capital made available, we would still be faced with the problem that, to date, there are only a relatively small number of specialists trained in renewable technologies. Decades of disdain for renewable energy have been accompanied by its neglect at all levels and in all subject areas of school, professional and academic education. Without an international educational initiative aimed at closing this gap, rapid energy change cannot be effected. Education takes time, but we cannot and must not wait to introduce widespread energy change until a new generation has been adequately educated. Today's young generation is more interested in renewable energy than in any other subject area. This is particularly the case for those aiming at a technical or scientific career and who wish to play a role in solving one of society's fundamental problems. Information technology has recently demonstrated how huge numbers of people can adapt to a fundamentally new technology in only a short space of time and on a global scale.

Craftspeople, engineers, farmers, foresters, physicists, chemists, biologists and architects are all needed in the field in renewable energy, as well as economists able to think in terms of ecology and social costs. After a warming up period of around a decade, a jolt has gone through the German universities, although this has been more the result of student demands for relevant courses than the efforts of science managers. Now almost every German university offers a master's degree in renewable energy. Many chambers of trade have begun to prepare their members by offering further education in renewable energy, and new training centres have been established. None of this happened by itself; rather it is the result of intensive campaigning at thousands of information events.

In its race against time, the key question for energy change is how to fill the gap between the current demand for specialists and the long time periods required to educate a new generation of specialists. This knowledge is needed now, and not in 10 or 20 years! The most obvious practical solution is widespread professional further and continuing education. It must also be open to those already working and unable to stop work in order to study. This is relatively easy for professional trade organizations to organize at regional level.

An international postgraduate university for renewable energy

Training must also be organized at an international level. The only practical approach for educating specialists at academic levels is by means of a virtual university: offering postgraduate studies via online teaching, so that anyone can take an additional degree in renewable energy from their own homes and without having to abandon their current careers. Many of the current generation of engineers, architects or chemists, aged between 30 and 50, are open to energy change, although having scarcely touched on the relevant technologies during their own university studies. EUROSOLAR and the World Council for Renewable Energy have been recommending the introduction of international postgraduate studies, in all languages, using e-learning and in the form of an international "Open University for Renewable Energy" (OPURE). It has not yet been realized, although it was once almost implemented. In my speech at UNESCO's World Solar Summit in Paris in June 1993, I proposed setting up an international university for renewable energy. The suggestion became the main topic of the *World Renewable Energy Agenda,* published in 2004 by EUROSOLAR and the World Council for Renewable Energy which was founded in 2001.[76] The German Federal Minister for Education and Research at the time, Edelgard Bulmahn, seized on this proposal and made it a central topic of her ministry's International Science Forum on Renewable Energy, part of the Renewables 2004 conference to which the German government invited UN governments in June 2004. In December 2004 the Bundestag, the Lower House of the German Parliament, welcomed this proposal in a resolution and demanded that the federal government "drive forward the establishment of an Open University for Renewable Energy (OPURE) and attract international partners to the project". During the spring of 2005, a memorandum of understanding on OPURE was signed between the United Nations Educational, Scientific and Cultural Organization (UNESCO), EUROSOLAR and the Institute of Solar Energy Distribution Technology (ISET). However, after the German Parliamentary elections in September 2005, which led to a change of minister at the Ministry of Education and Research, this initiative was allowed to fizzle out.

Today it is more necessary than ever. The increasing number of initiatives for renewable energy being undertaken in many countries worldwide has made the lack of specialists increasingly obvious. By enabling people with a technical and scientific background and professional experience to gain an additional qualification, it would be possible in just a short space of time to acquaint many people with the technological and economic characteristics and requirements of renewable energy and its applications. International educational standards could also be established in this manner.

The OPURE concept is practical, has a widespread, stimulating effect and is also a prerequisite for promoting the production of renewable energy technologies in developing countries. It costs decisively less – and has a decisively greater effect – than the control and certification mechanisms established at national and international levels for emissions trading. However, it does require that political attention is finally paid to adopting a practical and relevant focus, also at an international level, and that targets are clarified. This will help to speed up the decision-making process, whether via a "coalition of the willing" which takes up initiatives such as this, or sensibly charging UNESCO with its operation and providing the necessary funding. This would simultaneously provide one of the largest UN Special Agencies with a new task, itself a contribution to UN system reform.

Quo vadis IRENA?

The process of establishing an international government organization is lengthy. This also applies to the International Renewable Energy Agency (IRENA), established in January 2009 after many years of campaigning, and which started work in July 2009 after its location and management had been decided upon. The agency is the result of a EUROSOLAR initiative, based on a memorandum published in 1990. There are already international government organizations dealing with questions of energy, such as the International Atomic Energy Agency (IAEA) and Euratom, both founded in 1957, and the International Energy Agency (IEA), founded in 1974. However, these are exclusively focused on either the international promotion of nuclear power, or of all sources of energy – such as the IEA, which systematically talks down renewables. Therefore it was essential to have an international agency dedicated to renewable energy alone. IRENA's founding is closely linked to my own efforts, without which it would never have been established. Therefore it would be legitimate to regard my own evaluation of its achievements to date as biased. However, I want to recapitulate the process involved in the setting up of IRENA and its current activities in order to show the necessity of such an organization, as well as the resistance and problems it faces.

The path to establishing IRENA was long and uneven, and initially almost no one believed it realistic or possible to implement.[77] The start was promising. The initiative was taken up at UN headquarters, where I had presented the idea during the spring of 1990, and from the very beginning it was supported by the UN official then responsible for questions of energy, the former Foreign Minister of Mauritania, Ahmedou Ould-Abdallah. Javier Pérez de Cuéllar, then UN General Secretary, was persuaded of the value of the initiative, and forwarded it to the United Nations Solar Energy Group on Environment and Development (UNSEGED). This group was working up proposals for promoting renewable energy at international level, with a view to the upcoming Rio conference, and recommended establishing IRENA. However, this recommendation was rejected by a majority of the industrialized countries as well as the OPEC countries who were sitting on the preparatory committee for the Rio conference. The idea appeared to be dead. It was obvious that it could only be achieved by working outside the UN system, which is reliant upon consensus. Therefore, as nowhere is it written that an international organization must be a UN organization, our next step was to attract a government, get it to take up the idea and motivate a group of other countries to work with it and make IRENA a reality. After several attempts and more than ten years of intense effort, we succeeded in pushing through an initial political resolution: it was agreed in the 2002 programme of the Social Democrat/Green coalition that the German government would drive forward this international initiative. However, initially the federal government was not prepared to actually implement the resolution.

Quite the opposite in fact. At the German government's Renewables 2004 conference, originally conceived as a conference for motivating support for IRENA, the initiative was deliberately excluded from the agenda, it being feared that many countries would refuse to take part in Renewables 2004 if IRENA were made an official item. Thus it remained the job of the International Parliamentary Forum for Renewable Energy, run in parallel to the conference, to demand IRENA's establishment.[78] For inexplicable reasons, even international environmental organizations, including Greenpeace and the WWF, spoke out against this initiative. Only after my renewed efforts to have the IRENA initiative included within the programme of the subsequent German government (the CDU/CSU and Social Democrat coalition, formed in the autumn of 2005) was the process actually initiated in practice. During 2007 and 2008, three especially selected envoys and I undertook official discussions with more than 50 governments to encourage them to take part. During the founding conference, which was finally held in January 2009, 75 member countries signed the declaration of enrolment. In June 2009, in Sharm el-Sheikh in Egypt, the location of IRENA's official headquarters was agreed and its director-general elected. Three cities had been proposed by their governments as the headquarters for the organization – Abu Dhabi, Bonn and Vienna – and the decision was made to locate in Abu Dhabi, capital of the United Arab Emirates. By this time, IRENA already had 135 member countries. Hélène Pelosse, proposed by the French government, was elected as director general. I myself was not a candidate, although many had expected me to be so and I was a favourite, having driven forward the IRENA initiative for so many years. Many well-known international advocates for renewable energy had also supported the cause, including Robert F. Kennedy Jr., Amory Lovins, Bianca Jagger, David Suzuki, Jakob von Uexkuell and Ernst Ulrich von Weizsäcker. But the German government, with my agreement, did not nominate me, in order not to thwart its efforts to win Bonn as the future headquarters for IRENA.

Meanwhile, more than a year since IRENA started work, 147 countries have signed up. By July 2010, more than 25 parliaments had ratified the treaty so that the organization was now also constituted according to international law. However, it is disheartening to note that the agency is already bound by resolutions passed at the government conference of the member states in Sharm el-Sheikh, particularly in its financing from member states' contributions. Instead of a set contribution, proportional to the economic strength and population of each member country (the so-called UN key) an initial budget of only US$25 million was specified, so that with increasing numbers of member states, the contributions of each member state decreases proportionally. Yet involving more member states brings an increasing number of tasks and requires a larger budget, something this disastrous resolution prevents. This makes it impossible for IRENA to achieve its primary role of advising the governments of member states on renewable energy policy. In January 2010, the Administrative Committee, a supervisory body made up from representatives of member governments, set the budget for IRENA's first year at only US $13 million – a pathetic sum for 147 member states and insufficient even for it to start its intended activities. But even this sum was not actually made available, for initially most member governments (until the treaty was recognized by international law) hadn't paid their "voluntary" contributions, and even afterwards, there were delays. A lack of budget means that the initial work of the organization suffers because only very few staff members can be taken on.

Despite its founding, it is clear that the necessity of a powerful agency has not been understood. Several governments only joined up after it was no longer possible to stop IRENA becoming a reality. Although it is an unmatched achievement for an international governmental organization to bring together so many member countries during its founding (IRENA already has more member countries than the International Atomic Energy Agency, IAEA), with their lack of personnel and operative room to manoeuvre, it is cursed with feet of clay. The imbalance between the number of member countries and its minimal financing is intolerable. IRENA's operating budget is no larger than that of a single, medium-sized development aid project, and lies significantly below the annual membership contribution of €29 million that Germany alone makes to the IAEA. Failing an immediate, fundamental change in this situation, we will have missed an enormous opportunity to make IRENA an international instrument for supporting political energy change strategies.

When we compare the origin of IRENA with the genesis of the IAEA in 1957, then the differences are as glaring as they are paradoxical: the IAEA was jump-started in a very short space of time and was able to rely on the wide and enthusiastic support of all governments, the UN, and generous funding. Its foundation reflected the spirit of its times, the entering of a new age: the atomic age. In contrast, 50 years later, establishing IRENA has meant battling laboriously against the resistance of established international government organizations as well as the UN system itself and the World Bank, and even those governments who introduced the formal steps for its founding at international level initially had to be "carried to the fight". Governments showed little enthusiasm, as if renewable energy were a non-issue rather than the gateway to the solar age – our only means of fulfilling the promises so thoughtlessly and rashly made by the proponents of nuclear energy. This demonstrates the size of the gap between a commitment to renewable energy and the understanding that this commitment demands at least as much political effort as has been spent on nuclear power for decades.

Although within our grasp, if we fail to enter the solar age it will not be the fault of the "subject" itself, but rather of political faintheartedness and short-sighted interests. It may also be due to the power of international policy-making (once used to drive forward nuclear energy, even overcoming political divisions between "East" and "West" to do so) having been exhausted in an excess of global conferences, themselves increasingly a form of marshalling yard.

D. WINDING UP THE ATOMIC AGE: PHASING OUT NUCLEAR ENERGY BY MEANS OF A GLOBAL BAN ON NUCLEAR WEAPONS

The "peaceful use" of nuclear energy achieved popularity because it was seen as an alternative to atomic weaponry and nuclear war. But in reality, atomic armaments have become the final anchor for nuclear power. Having nuclear weapons still remains a sign of "first-class membership" on the world stage and, with the five permanent members of the UN Security Council all nuclear powers, is explicit proof of global standing. However, the five official nuclear powers (together with Israel, an unofficial nuclear power) of 1990 – a historical turning point – have now become nine: after their spectacular nuclear weapons testing in 1998, Pakistan and India have joined the group, as has the rogue state of North Korea. Iraq wished to join the league of nuclear powers but was forced to put aside these ambitions after the Gulf War in 1991. It is possible that Iran is attempting to become a member. The Ukraine and Kazakhstan, Soviet Union successor states, are both technically able to build nuclear weapons. Prior to the end of the apartheid regime, the Republic of South Africa had also considered taking this step. Perhaps they will return to these plans at some time, for owning nuclear weapons, even if they are never intended for use, definitely raises a state's political standing in the global political world-order. Other countries such as North Korea, and perhaps soon also Myanmar (Burma), want to use nuclear weapons as a political pawn, ensuring deference from the international community who will, when in doubt, turn a blind eye to their human rights violations. All in all, the motivation for possessing nuclear weapons has increased. Nine nuclear powers now – and perhaps 12 to 15 by the year 2020?

Clearly, existing nuclear weaponry or its aspired possession cannot be seen separately from the question of nuclear power. No state which owns and wishes to retain nuclear weapons (and none who is secretly striving for nuclear weapons or, without the knowledge of its own population, wants to keep this option open) will be willing to give up its own nuclear power plants. If you have, or want, atomic bombs, not only do you need nuclear power plants, you also need the basis for an atomic technology industry. For every nuclear power, nuclear technology is a "double-use technology": having nuclear weaponry without one's own atomic technological potential is unthinkable, and maintaining such a potential solely to build nuclear weapons is almost unaffordable. Thus for as long as we have nuclear weapons, attempts will be made to stimulate a "renaissance in nuclear power". But no government will admit to holding on to its nuclear power plants simply to maintain this status. Instead, together with the atomic energy organizations, nuclear powers desperately seek justification for arguing that renewable energy alone is insufficient to meet energy demands. And this is how excellent nuclear scientific knowledge comes to be paired with ignorant arguments against renewable energy. Putting a stop to nuclear energy means nuclear disarmament, otherwise there will be ever greater and more influential attempts to limit renewable energy. Governments that recognize and work towards the target of using renewable energy to meet all their energy needs must also accept the goal of nuclear disarmament. Any other path would be inconsistent or blind to the true circumstances.

It is the five established nuclear powers who bear chief responsibility for adherence to nuclear armaments. It is they who, for years, have contravened the 1970 Nuclear Non-Proliferation Treaty. This treaty has three main pillars: in Article IV, all member states commit themselves to the "fullest possible exchange of equipment, materials and scientific and technological information for the peaceful uses of nuclear energy". This was the spirit of period from the 1950s to the 1970s, where nuclear energy was regarded as *the* energy source of the future. In Article II, each signatory to the treaty who does not yet possess nuclear weapons commits not to seek to arm themselves with nuclear weapons in the future and "undertakes not to receive, from any source, nuclear weapons, or other nuclear explosive devices and not to manufacture or acquire such weapons or devices". And in Article VI, the nuclear powers commit to negotiating complete and controlled nuclear disarmament.

During the 1970s and 1980s, international contraventions of these obligations were tolerantly ignored – the world was waiting for nuclear disarmament negotiations between "East" and "West". With the end of the East-West conflict and the disintegration of the Warsaw Pact, it appeared that the opportunity for worldwide controlled and complete nuclear disarmament had finally arrived. But the West ignored this opportunity. The North Atlantic Treaty Organization (NATO) maintained its nuclear strategy because of "new dangers in the south" (although there are no nuclear weapons in this region), thereby indirectly opening up the next, extended round of nuclear armament. Indignation at India and Pakistan's atomic tests in 1998 (officially setting the seal on their membership of the club of nuclear powers) was hypocritical, for the critics' adherence to their own nuclear weapons is equally scandalous. International sanctions against Iran, imposed because of its nuclear ambitions, have no credibility and are without moral authority so long as the nuclear powers themselves hold onto their own nuclear weapons, thereby making their own contravention of the treaty clear.

In September 1968, as the Non-Proliferation Treaty was being negotiated, a conference of the "Non-Nuclear States" was held in Geneva. This conference pushed through Article VI, without which the treaty would never have come about.[79] Today all these "have not states" should come together again to exert combined pressure on the nuclear powers, forcing them finally to fulfil the conditions of Article VI on complete global nuclear disarmament. Only when this happens will there be sufficient global political legitimacy for taking massive steps against any state attempting to arm itself with nuclear weapons. In the meantime, even prominent American advocates of nuclear deterrence, such as former foreign Secretaries of State George Shultz and Henry Kissinger, have recognized the need to do so. Together with the former US Secretary of Defense, William Perry, and the former chairman of the US Senate's Defense Committee, Sam Nunn, they wrote a spectacular article in *The Wall Street Journal* in 2008 arguing for complete global nuclear disarmament, in view of the latest activities in Iran. In his equally spectacular speech, given in Prague in April 2009, newly-elected US President Barack Obama demanded the same. But where is the massive international support for this initiative? Where are the initiatives from Europe's nuclear powers, France and Great Britain, and why aren't the non-nuclear states joining together to demand action? Only with complete nuclear disarmament can we close the chapter of nuclear power and end the atomic age – *the* greatest global aberration of the second half of the 20th century, and the barrier to entering the solar age.

Eliminating nuclear power

The scale of our nuclear aberration is demonstrated by the legacy of atomic waste which will be with us for a historically unique period of time. Ethnologists talk of 10,000 years of human civilization being confronted with hundreds of thousands of years of radioactive nuclear waste, produced within a period of only 65 years. This alone makes discussions about the ethics of nuclear power unworthy: even if the nuclear waste is pushed underground, it cannot be securely stored over the long term as it is anything but "dead material". Burying it is simply a case of "out of sight, out of mind", before its radioactivity presents us with insoluble problems. Thus a part of the winding up of the "peaceful use" of nuclear power involves asking the question whether it is in any way responsible to consider the ostensibly "permanent disposal" of nuclear waste. The International Nuclear Fuel Cycle Evaluation (INFCE) conference initiated by US President Carter in 1979 considered this question, as did – also in 1979 – the Commission of Enquiry into the future of nuclear power, set up by the German Bundestag under the chairmanship of the Social Democrat parliamentarian Reinhard Ueberhorst.

The Commission examined the concept of constructing one or more "ex-territorial" islands, for example in the Pacific Ocean, designed exclusively for the permanent disposal of nuclear waste. It also considered the concept of engineered storage, in the form of above-ground storage systems, constantly monitored, so that the nuclear waste remains accessible for treatment or transfer to a permanent storage facility at a later point.[80] What we now call "temporary storage" will eventually become permanent disposal sites – and ideally geographically close to the nuclear power plants themselves. Governments wishing to hold onto nuclear power will have to overcome the embarrassing contradiction of wanting to operate their own nuclear power plants while transferring the burden of nuclear waste to someone else. This includes the proposal for *transuran* research, as advocated by the winner of the Nobel Prize in Physics, Carlo Rubbia. Its aim is the transmutation of nuclear waste, whereby radioactive isotopes with very long half-lives are transformed into more short-lived ones, so that they lose their radioactive radiation within a manageable period of time, and with the final product being stored in permanent disposal facilities. However, we do not know if this process, based on the use of particle accelerators, actually works. It will take decades to find out.

In any event, it would require significant quantities of energy. It probably requires more energy to transmute nuclear waste than it originally contributed to our energy needs. Effectively this would be a debt that we pass to future generations, one that can only be paid off by the complete transition to renewable energy. This is the only form of nuclear research which remains legitimate. However, it could only be enforced in conjunction with the definitive decision to decommission nuclear power plants. But it would be the only responsible means of disposing of nuclear waste. Our legacy of nuclear waste demonstrates just how presumptuous our attempts at an "atomic age" truly were.

Highly qualified nuclear physicists, chemists and engineers must take over the long-term task of phasing out nuclear power. They also need to be highly paid, to persuade them to dedicate their careers to clearing up after the mistakes of the past. Even then, only those with an extreme awareness of their ethical responsibility to the environment will be prepared to undertake such a task. The dilemma of our civilization, a dilemma to which the Promethean experiment of nuclear energy has led us, makes it very clear: we cannot allow the continued growth of nuclear waste, we must stop using nuclear power, and we must carry out the controlled disarmament of nuclear weapons which are the ultimate reason for retaining nuclear power plants. If not now, then when?

7

A VALUE DECISION

Social ethics instead of energy economism

In July 2010, the UN General Assembly declared that access to clean water was a human right. Although this is undoubtedly a basic right, this resolution does not yet enable individuals to enforce it directly. If or when that becomes the case, it would have far-reaching consequences, even prohibiting production methods that pollute waters and thereby directly endanger human health. Human rights are a form of ethical principle and we cannot allow adherence to these principles to be dependent upon whether this "adds up", or is a threat to "competitiveness".

The current UN General Assembly resolution on the human right to clean water is the result of increasing efforts to define human rights more strongly and preferably to make them enforceable, to award them priority over general rights to freedom and equality, and to incorporate social rights, such as the conservation of nature.

In light of the worldwide social crisis, one inextricably linked with the ecological crisis, the scope of human rights takes on new dimensions. The clause in Article 2 of Germany's Basic Law, in which every person has the right to "physical integrity", had long been primarily understood as a prohibition on physical violence against individuals and protection from physical injury. But even within the terms of the German constitution, environmental contamination is a far greater threat to physical integrity. Article 1 of France's 2004 environmental charter, part of its constitution, states that "everyone has the right to live in a balanced, healthy environment". In the EU's Charter of Fundamental Rights, an integral part of the EU Treaty, Article 37 states that "a high level of environmental protection and the improvement of the quality of the environment must be integrated into the policies of the Union and ensured in accordance with the principle of sustainable development". However, so long as we adhere to the established power system, not only is it impossible to make these fundamental rights an everyday reality, it is also impossible to implement the newly declared human right to clean water or the much-discussed but not yet officially declared human right to clean air, available energy or to truly sustainable (i.e. for coming generations) methods of cultivation. These rights can only be realized through the transition to renewable energy, and thus for human rights reasons there is a political obligation to act.[81]

The transition to renewables is technically possible, thus removing any ethical justification for its delay. Not even valid economic objections remain. Claudia Kemfert, head of the department of Energy, Transportation and Environment at the German Institute for Economic Research (DIW), calculates that, by the year 2015, Germany alone will be facing costs of €50 billion to remedy environmental damage, €10 billion for initial adaptive investments and €40 billion to cover the increasing costs of fossil energy. By the year 2025, these costs will have increased to €290 billion.[82] A report published by the UN in July 2010 on the environmental damage caused by the world's 3,000 largest companies comes to the economically dramatic conclusion that, by abusing natural resources (above all through emissions of greenhouse gases, other exhaust gases and water pollution), these companies cause damage amounting to US$2 billion each year. Extending this damage evaluation to all economic areas, and to all the damage caused by fossil and nuclear power supply, indicates far greater costs and the misguided use of economic resources. This calculation includes people who die mining coal or uranium, or those who develop leukaemia from living close to a nuclear power plant, the public and hidden subsidies for nuclear and fossil energy amounting to hundreds of billions of dollars each year, through to the political and military costs of "international energy security" which costs the US alone hundreds of billions each year, as Amory Lovins and his co-authors of *Winning the Oil Endgame* calculate.[83]

An end to excuses

All this continues in spite of the general economic calculations set out here, which demonstrate that energy change would cost not more, but increasingly less, than continuing with established energy supply systems. The social as well as economic advantages of a transition to renewables can no longer be seriously disputed. Quite the opposite: current investments into renewable energy are the precondition for safeguarding the supply of environmentally-friendly, cheap and sufficient energy for everyone, now and for all future generations. It is the historic responsibility of the current generation to make this transition on behalf of future generations. There are no more excuses. All the obstacles encountered during this process are easier to overcome than the consequences of carrying on as we are. It is a huge socio-psychological error to assume that it first requires a catastrophe to strengthen the will for change and to improve the chance of its being realized. Enormous efforts can only be undertaken by societies which are sufficiently stable and not subject to a state of emergency.

I have illustrated in detail in this book the reasons why this process must not be entrusted to the established power industry. Their sole interest is to slow down the process of change and to deploy renewable energy in a technologically lopsided and suboptimal manner – a deployment in the rational economic interest of large companies but not in the national economic and social interest. Rather than "centralization versus decentralization", energy change can be better described as the transition from a system of energy supply organized in large-scale, interdependent networks, to one which is autonomous and highly modularized – individual, local, regional, and each over a range of scales. Thus we are not dealing exclusively with a contrast between small and large plants – where there is a large demand for energy, e.g. for major companies, municipalities or regions, then correspondingly large-scale plants can supply this energy using a modular or insular structure.

It is true that it would be conceivable to have a renewables-based power supply system based on a series of decentralized production plants which are linked together into a large-scale international power supply network. However, unlike conventional energy, the nature of renewable energy renders this form of organization unnecessary. Apart from the mental habit of thinking in terms of conventional energy supply structures, there is no reason to take this route. It would be a long and complicated method, inevitably involving dependencies, of organizing something which can be effected in a short, simple and independent manner. Thus it is this tension between autonomous and varied forms of production and use on the one hand, and structural dependency on the other, which characterizes not only the conflict between renewable and conventional energies but also discussions on the design and structure of a future energy supply system based exclusively on renewables. The concept which wins through, and goes on to influence preliminary political decisions, will also decide whether the process of energy change is speeded up or slowed down. It has been empirically proven that modular structures offer the greatest potential for acceleration.

The concept of a central, internationally-managed network, linked with a power trading exchange which regulates demand and thus production quantities and prices for major producers as well as innumerable networked producers, is only convincing in theory. Supplying power in this way would make much that is currently independent and transparent once again dependent and opaque. An apparently high degree of economic rationality would effectively result in an almost intolerable situation. It is not by chance that the computer age, with its culture of applications, contradicts this apparent rationality. It is inconceivable that the capacity represented by the hundreds of millions of laptops could be planned, let alone that such planning would be accepted. The mass use of laptops also represents an almost incalculable overcapacity, one which, in economic terms, is highly inefficient. It is all the more creative and natural for being so. It is not only economic motives which decide whether an autonomous and practical service technology takes off. A development in renewable energies, similar to the revolution in information technologies, is looming.

Value synthesis

Making renewable energy available in an increasingly autonomous and democratically controllable manner renders possible hitherto undreamt-of value syntheses:

- between individualism and the common good, the classic philosophic theme, because its autonomous use expands individual freedoms without burdening others;
- between spiritual and material values, because the material interests of mankind can be satisfied without social and environmental damage, thereby creating an ecological economy.

This explains the growing popularity of renewable energy, popularity based on the (perhaps only intuitive) recognition that these possibilities exist. Citizens have recognized the fundamental potential of renewables to a far greater extent than most governments realize and the traditional power industry wishes to believe. This is highlighted by a survey from the UN Report noted above. Eighty per cent of consumers value ecological methods of production and would favour market restrictions, for they are not solely – as the neo-liberal model claims – motivated by the short-termism of "homo oeconomicus". Another survey on the German population's attitude to renewable energy is even more convincing – according to the survey, 75 per cent of Germans favour the complete transition to renewable energy, 48 per cent believe this is already possible. 74 per cent want to maintain the current level of subsidies, 61 per cent favour communally-owned power plants, with 58 per cent of them willing to invest in them. Of those questioned, 82 per cent wanted external costs to be included in energy price comparisons and 88 per cent wanted these costs be stated in energy bills, while 76 per cent believe these costs should be carried by the operators. But only 19 per cent are satisfied with the efforts of state and municipal politicians in driving forward renewable energy, with 91 per cent demanding they act more forcefully.[84]

The social movement towards renewable energy is based on a variety of motives which, despite its members' differing values, interests or political orientation, must be seen together, even though a single reason is sufficient to encourage involvement: global climate protection, immediate quality-of-life, technological innovation, new economic perspectives, self-determination and the democratization of living standards. One of these motives alone is insufficient to stimulate a whole movement: a movement is created by that which I call the "socio-logic of renewable energy".

The greatest mistake that advocates of renewable energy can make is to get drawn into purely economic discussions of energy which reduce the debate to comparisons of current prices. What is decisive for energy change is the *social importance and vision of renewable energy* – and not the opportunist political striving for energy consensus, the standpoint of the traditional power industry or of companies in the emerging renewable energy technologies sector who, as they develop, increasingly and inevitably split into individual interests in the effort to become "normal" companies. Of course, for these companies and their employees, there is a fundamental ethical difference between producing renewable rather than nuclear or fossil technologies. But like every other company, they are obliged to make economic decisions. Discussions on renewable energy which simply compare the "solar industry" sector with the traditional power industry end up considering energy in purely economic terms, and neglect the social value of renewable energy.

Seen from the standpoint of economic reductionism, every campaign against renewable energy, instigated by the traditional power industry via the energy institutes which they sponsor, is designed to deflect attention from that industry's own problems and the key questions of energy production and supply. In September 2010 the German Association of Energy and Water Industries (BDEW) started a campaign against Germany's Renewable Energy Sources Act (EEG). Its slogans hinted at the cost of subsidizing the widespread introduction of renewable energy, a cost borne by all energy consumers. It asked: "How much solar subsidy is enough? How much is too much? When and how can it be integrated into the market? Do we need alternatives to alternative energy? And what are the answers? Germany is ready for the debate on energy." Other campaign slogans included: "In Germany it's only dark when you want it to be. In the USA it's dark for 144 minutes a year, whether you want it or not," and "Energy isn't black-and-white." These statements were designed to suggest that it is "better to accept all the nuclear power and climate risks and their consequences than to suffer power cuts lasting a couple of minutes which could result from the use of renewable energy! The share of renewable energy is getting too large, and it must be introduced on our terms! The alternative to alternative energies is what we do! Germany is ready to hold back the expansion of renewable energy! We want to be the recipient of these subsidies."

Considering the average annual profits of €20 billion racked up by Germany's four major power companies and the social costs they generate, campaigns such as these are impertinent. They make an artificial distinction between energy consumers and citizens although, without exception, every energy consumer is a citizen who, through taxes and insurance contributions, is obliged to pay for all the social costs which do not appear on their energy bills. Campaigns of this type aim at stripping renewable energy of its public legitimacy and helping traditional power companies win back their dominant role in directing energy discussions. They can only succeed by reducing the debate about renewable energy to one about prices. Their goal is to slow down the process of energy change and be in charge of it, in order to nullify the social movement towards renewable energy.

The renewable energy movement thrives because it has not suffered a setback and its motives remain valid. A second start is always more difficult than the first. Developments in Germany since the 1990s, ignited primarily by the EEG, have stimulated and inspired the international dawn of renewable energy. Thus the conflict over renewable energy is of global significance. Any temporary backlash could set back the international movement.

System decision

Having begun to accelerate its own pace of change, Germany has made itself the arena for structural conflict, for here the decision on systems is pending. During the period in which the Social Democrat/Green coalition was in power – the period in which the transition to renewable energy took its first steps – the German government avoided making this decision. Both energy systems were promoted and strengthened equally, with the result that we now have two trains on the same track, rushing at one another head-on. On the one side we have the EEG, a strongly developed programme of market incentives for renewable energy, a cogeneration Act (KWKG), a programme for increasing the energy efficiency of old buildings, a meagre eco-tax and the Act on the structured phase-out of nuclear power for the production of electricity. On the other, the process of establishing a control body to oversee the legally decreed liberalization of the electricity and gas market has been dragged out for years, the energy industry's process of concentration has been supported, and a law on emissions trading has been drafted that increases power company profits by many billions. A political price had to be paid for the nuclear power phase-out Act, a price known as "consensus". This consensus took the form of "peace obligations" in which the economic privileges of the nuclear industry remain untouched. They also include tax-free reserves, the waiving of a tax on nuclear fuels and the requirement for third-party insurance, representing annual bonuses to the industry of around €6 billion. To date, the effective result of introducing the nuclear phase-out Act has been that power companies have pocketed the billions given as the political price for consensus, using the funds to further concentrate their operations. At the same time they are betting that the new German government will extend the lifetime of nuclear power plants without demanding anything, i.e. decommissioning, in return.

The "energy policy plea", published in August 2010 in the form of full-page announcements in all the German daily newspapers, mirrors the attitude of the traditional energy industry. In language which reveals an overweening and indefensible claim to the position of exclusive agency and font of all knowledge on questions of energy, the announcements consist of a chain of apparently self-evident and suggestive statements.

We read: "Take up the challenge: the future belongs to renewables," but with the immediate corollary "and CO_2-free energies" – which is an attempt to smuggle nuclear power and CCS power plants into the renewable energy camp under the guise of " future energies". The production locations for renewable energy are also predetermined: "Wind power comes from the North Sea and Baltic Sea, solar power from Southern Europe and perhaps even one day the Sahara." Decentralized and everywhere? That can't be allowed!

The text continues: "Enormous investments are needed to expand renewables. The necessary financing must be generated by the energy suppliers" – as if they were the only ones making relevant investments in renewable energy. Just like eco-taxes, the tax on nuclear fuels suggested by the German government would, it claimed, "hinder future investment". In contrast, the continued use of cheap nuclear and fossil fuels would be a faster route to renewable energies, because the power companies would anyway be using their nuclear and fossil profits to concentrate solely on renewables.

A little further on, it states: "Many of the new energies will be produced far away from the points of consumption in Western and Southern Germany." Thus "we must urgently develop and construct new, efficient and intelligent power networks and energy storage solutions" which demands "less bureaucracy and faster authorizations". This is an indirect vote for the super-grid, for nowhere do they mention that most renewable energies can, instead, be produced directly at the point of consumption. (This is the quicker and more cost efficient solution, requiring only the long overdue reduction in bureaucracy and faster authorizations to make it reality.)

And finally they conclude: "The regenerative energy turnaround cannot be achieved overnight. Renewables need strong and flexible partners. These include modern coal-fired power stations. They also include nuclear power. a premature phasing-out of nuclear power would wipe out capital to the tune of many billions – burdening the environment, the economy and the population of our country." For the authors of the "energy policy plea", the social and ecological costs of nuclear and fossil fuels are not worth mentioning. They follow the path that has already led world civilization into an existential vicious circle – demanding more environmentally damaging growth in order to generate sufficient funds to repair the damage caused by this growth. An abstruse form of logic!

Those wishing to sideline the new technologies which will stimulate economic regeneration while the old technologies are still operable, are preventing this very economic regeneration from happening. Joseph Schumpeter, one of the economic greats of the 20th century, speaks of the necessity of "creative destruction", to make way for economic innovations. If it were up to the "energy policy plea" signatories from the four major German power companies, then the established power industry would obviously be exempted from this requirement. The signatories include the head of Deutsche Bank and the chairmen of the boards of BASF, Bayer and ThyssenKrupp. They don't even notice what these demands for protection would mean if they were extended to other industrial sectors: a government would be obliged to ensure that all companies use the capacity of all their available production facilities to the full and keep their competitors out of the market, to ensure that no investment capital is destroyed. This is an absurd attempt, which takes a planned economy as its model, to delay independent investments into renewable energy in the name of business and society, and to promise society that its much-needed clean energy will come in the future.

Energy change itself is unavoidable because nuclear and fossil fuel resources will become exhausted, that is to say, they will reach their inherent limits, and before this happens the external costs of these fuels will be felt. However, rapid energy change unavoidably involves system change. Due to the nature of renewable energy technologies, this is pre-programmed. Yet system change cannot be allowed to be indiscriminately held back. This would slow down and delay energy change, putting it into the hands of a few. For too long we have accepted the major energy institutions' claims of primacy when it comes to competency in dealing with matters of energy supply, and these institutions have been happy to stand up on behalf of "the economy", awarding themselves their own political mandate to act. Thus the precondition for rapid energy change is the mental and practical emancipation of society, its technological industries and political institutions from the traditional energy system. Only then can the unfolding technological emancipation from conventional energy structures develop fully. It is of unparalleled historical importance that this transition is executed now, and with the involvement of many, and not a whole generation of people and power stations later.

The system change we need to effect now is thus part of the "ethical imperative" of which philosopher Peter Sloterdijk writes in his book *Du mußt dein Leben ändern* (*You Must Change Your Life*), urging that this is not something to be taken lightly. "The sorcerer's apprentices of planetary design have been forced to realize that the incalculable is a whole dimension ahead of strategic calculations." Thus there is also "no right to face only those problems that we can solve using the means at hand".[85] Thus the rapid transition to 100 per cent renewable energy will be achieved by multiplying the number of activists, each with their own motives, who are not subject to the systemic logic of the traditional energy regime. The most important political maxim is to open up and continually expand the necessary space in which they can operate. The course must be set by implementing two decisive measures:

The first is to turn the social and economic advantages of renewable energy into individual economic incentives, such as those introduced by the EEG. The best way of doing so would be to subject the entire energy supply system to an emissions tax, coupled with the principle of market priority for renewable energy. The second is to implement the overriding principle of priority for renewable energy in land-use planning, in order to clear away anachronistic bureaucratic obstacles. Society's own economic power will take care of everything else almost automatically, especially at municipal level and once the relevant technologies are mass produced. There is no need for an "overall energy policy" to be agreed with the power companies, with compromises that include nuclear power and coal-fired power stations. Rather, we need policies which encourage innumerable investments into energy change. The question of whether society's transition to renewable energy will be accelerated is one that will be decided by politics. The *energy imperative* is – full speed ahead.

NOTES

1 www.unendlich-viel-energie.de, 19 February 2010
2 Toralf Staud, *Grün, grün, grün ist alles, was wir kaufen. Lügen, bis das Image stimmt* (Cologne 2009)
3 Wilhelm Ostwald, *Der energetische Imperativ* (Leipzig 1912) p81ff
4 Erik Hau, *Windkraftanlagen* (Heidelberg 1989) p22ff
5 Marcel Perrot, *La houille d'or* (Paris 1962)
6 Wolfgang Palz, *Power for the World. The Emergence of Electricity from the Sun* (Singapore 2010)
7 Klaus Fuhrmann, 'Ohne Sonne geht gar nichts. Warum konventionelle Energiequellen nur marginal die Energiebedürfnisse befriedigen', *Solarzeitalter* (no. 2 2001) p36
8 Additional evidence of judgements of error can be found in: German Renewable Energies Agency, *Vorhersage und Wirklichkeit* (Kurzgutachten 2009)
9 In 2011, the German government decided to phase out nuclear power entirely, and combine this with an initiative for renewable energy
10 DEWI, *Status der Windenergienutzung in Deutschland*, 31 December 2009
11 A synopsis of these scenarios is given in my book *Energy Autonomy* (London: Earthscan 2007) on pages 60/61
12 European Climate Foundation, *Roadmap 2050: A practical guide to a prosperous low carbon Europe* (Brussels 2010)
13 Sachverständigenrat für Umweltfragen (SRU), *100% erneuerbare Energien bis 2050* (May 2010)
14 Renewable Energy Research Association (FVEE), *Energiekonzept 2050 – Eine Vision für ein nachhaltiges Energiekonzept auf Basis von Energieeffizienz und 100% erneuerbaren Energien* (June 2010)
15 Federal Environment Agency, *Energieziel 2050 – 100% Strom aus erneuerbarer Energie,* Authors: Thomas Klaus, Carla Vollmer, Kathrin Werner, Harry Lehmann, Klaus Müschen (July 2010)
16 Peter Droege (ed.) *100% Renewable Energy, Energy Autonomy in Action* (London 2009)
17 Michael Stöhr et al., *Auf dem Weg zur 100%-Region. Handbuch für eine nachhaltige Energieversorgung in Regionen* (Munich 2006)
18 Mark Z. Jacobson and Mark A. Delucchi, 'A Plan for a Sustainable Future: How to get all energy from wind, water and solar power by 2030', *Scientific American* (November 2009)
19 Greenpeace, *energy (r)evolution: A Sustainable World Energy Outlook* (2010)
20 Al Gore, *Our Choice – A Plan to Solve the Climate Crisis* (November 2009)
21 Lester Brown, *Plan B 2.0* (New York 2006) p254ff
22 Tamra Gilbertson and Oscar Reyes, *Globaler Emissionshandel – Wie Luftverschmutzer belohnt werden* (Frankfurt 2010)

23 Elmar Altvater and Achim Brunnengräber, 'Mit dem Markt gegen die Klima – katastrophe?', in *Ablasshandel gegen Klimawandel?* (Hamburg 2008) p10ff
24 Manuel Bogner, 'RWE's wundersame Klimaprojekte', *zeozwei*, (no. 2 2010) p36ff
25 Taken from the study by Uwe Leprich, *Stromwatch 2 – Die vier deutschen Energiekonzerne* (Saarbrücken 2009) and Bernd Wenzel and Joachim Nitsch, *Langfristszenarien und Strategie für den Ausbau erneuerbarer Energien in Deutschland* (Berlin 2010)
26 Ökonomisch höchst ineffizient, interview in the news magazine *Der Spiegel* (no. 50 2009) p60
27 Karin Holm-Müller, *Plädoyer für eine instrumentelle Flankierung des Emissionshandels im Elektrizitätssektor,* German Advisory Council on the Environment (SRU) (June 2010)
28 German Advisory Council on the Environment (SRU) *Weichenstellungen für eine nachhaltige Stromversorgung* (May 2009)
29 Gottfried Rössle, *Das Maren-Modell* (Hof 1989)
30 Öko-Institut e.V., *Analyse des Bedrohungspotenzials "gezielter Flugzeugabsturz" am Beispiel der Anlage Biblis A* (Darmstadt 2007)
31 Wolfgang Stieler, 'Sanfter Brüter', *Technology Review* (October 2009)
32 Physiker fordern Neubau von Atomkraftwerken *Spiegel online* (17 March 2010)
33 Mycle Schneider et al., *Der Welt-Statusreport Atomenergie 2009* (Paris and Berlin 2009)
34 Citigroup Global Markets, *New Nuclear – The Economics Say No* (November 2009)
35 Versuchsballon zum Klimaschutz: WWF begrüßt weitere Erforschung von Kohlendioxidabtrennung und Speicherung. www.wwf.de
36 Ulf Bossel, 'Carbon Capture and Storage. Aber wohin mit dem CO_2?', *Solarzeitalter* (no. 3, 2009) p20ff
37 Vattenfall, *Innovations- und Klimaschutztechnologien Carbon, Capture and Storage* (February 2009) p22
38 Reinhard Wolff, Norwegen landet doch nicht auf dem Mond, in wirklimaretter.de
39 IEA, *Energy Technology, Roadmap Carbon Capture and Storage* (Paris 2009)
40 David Hawkins and George Peridas, 'No Time Like the Present', Natural Resources Defense Council Brief (March 2007)
41 See note 37
42 Announcements documented in Deutscher-Braunkohlen-Industrie-Verein, Die Braunkohle. Was liegt näher? (2010)
43 Vattenfall, Innovations- und Klimaschutztechnologien Carbon, Capture and Storage (February 2009) p20
44 Marco Bülow, *Wir Abnicker. Über Macht und Ohnmacht der Volksvertreter* (Berlin 2010) p155ff
45 Johannes Frenzel, *Algen sollen Braunkohle-Treibhausgas fressen* Associated Press (July 2010)
46 Nina Scheer, *Welthandelsfreiheit vor Umweltschutz?* (Bochum 2008)
47 Olav Hohmeyer, *Vergleich externer Kosten der Stromerzeugung in Bezug auf das Erneuerbare-Energien-Gesetz*, Report for the Federal Environment Agency (2001)
48 Gert Apfelstedt, 'Vorrangregelung für Ökostrom unterm Damoklesschwert', *Zeitschrift für Neues Energierecht* (no. 1, 1997) p3ff
49 Hanne May, 'Attraktion mit langer Tradition. Wie Windräder, Naturschutz und Touristen Freundschaft schließen', in Franz Alt and Hermann Scheer, *Wind des Wandels* (Bochum 2007) p139ff
50 German Renewable Energies Agency, *Projekte in Kommunen* (Berlin 2008) p9
51 Fritz Vorholz, Die Strom-Offensive *Die Zeit* (29 April 2010)
52 deENET (ed.) *100% erneuerbare Energie Region* (Kassel 2009)
53 Eddie O'Connor, 'We can do it right now', *new energy* (no. 3, 2009) p33

54 *Energie und Management* (June 2010) p31
55 Hendrik Paulitz, 'Für eine kriegs-präventive dezentrale Energiewirtschaft in Bürgerhand', *Solarzeitalter* (no. 2, 2010) p3ff
56 Hartmut Rosa, *Beschleunigung- Die Veränderung der Zeitstrukturen in der Moderne* (Frankfurt 2005) p428ff
57 Helmut Tributsch, *Erde, wohin gehst du? Solare Bionik-Strategie: Energie-Zukunft nach dem Vorbild der Natur* (Aachen 2009)
58 Amory B. Lovins, *Small is Profitable* (Rocky Mountains Institute: Boulder, Snowmess, Colorado 2003)
59 EUROSOLAR: IRES, *www.eurosolar.org* (July 2010)
60 Thomas Dinwoodie, *Price Cross-Over Photovoltaics vs. Traditional Generation* (Sun Power Corporation Systems 2008)
61 Timon Gremmels, Raumordnungspolitik als Schlüssel zum Ausbau erneuerbarer Energien *Solarzeitalter* (no. 2, 2010) p16ff
62 Martin Unfried, 'Die Energieallee A7 – größer denken, offensiver kommunizieren', *Solarzeitalter* (no. 2, 2010) p21ff
63 Elinor Ostrom, *Governing the Commons* (Cambridge 1990)
64 Fabio Longo, *Neue örtliche Energieversorgung als kommunale Aufgabe* (Baden-Baden 2010) pp173, 348ff
65 Hermann Scheer, Mehr Tempo für Elektromobilität. *www.eurosolar.org* (July 2010)
66 Ulrich Grober, *Die Entdeckung der Nachhaltigkeit* (Munich 2010) p264ff
67 Otfried Höffe, *Democracy in an Age of Globalisation* (Springer 2007)
68 Thomas Fischermann and Petra Pinzler, 'Die Illusion von der einen Welt', *Die Zeit* (31 December 2009)
69 Eric Bruse, 'EU-Klimapolitik: Brüssel zieht Emissionshandel in Zweifel', *Handelsblatt* (20 May 2010)
70 German Advisory Council on Global Change (WBGU), *Climate Policy Post-Copenhagen: A Three-Level Strategy for Success* (Berlin 2010)
71 'On the scientific background to 350ppm' see www.350ppm.org (July 2010)
72 WWF Discussion Paper: Policy approaches and positive incentives for reducing emissions from deforestation and forest degradation (REDD) (August 2008)
73 L. Frenz, 'Amazoniens schwarze Sensation', *GEO: Das neue Bild der Erde* (no. 3, 2009)
74 Hans-Josef Fell, *Corporate Finance and Climate Protection: A Beneficial Alliance.* (Ms 2010)
75 Felix und Freunde, *Baum für Baum. Jetzt retten Kinder die Welt* (Munich 2009)
76 www.eurosolar.org
77 EUROSOLAR and WCRE: *The Long Road to IRENA. From the Idea to the Foundation of the International Renewable Energy Agency* (Bochum 2009)
78 German Bundestag: International Parliamentary Forum on Renewable Energies. Conference Report (Berlin 2004) p240ff
79 Die Konferenz der Nichtkernwaffenstaaten in Genf: *Europa-Archiv*, issue 21/1968
80 Hermann Scheer & Reinhard Ueberhorst: Wider eine irrationale Entsorgungspolitik, *Solarzeitalter* 3/2007
81 Nina Scheer: Vorrang für erneuerbare Energien? Chancen und Barrieren, politische und ethische Bewertung. *Amos international. Gesellschaft gerecht gestalten.* (Sozialinstitut Kommende Dortmund) 1/2010, p21ff
82 Claudia Kemfert, *Die andere Klima-Zukunft* (Hamburg 2008) p72f
83 Amory B. Lovins et al., *Winning the Oil Endgame* (Rocky Mountains Institute, Snowmess, Colorado 2004)
84 www.unendlich-viel-energie.de (1.7.2010)
85 Peter Sloterdijk, *Du mußt dein Leben ändern. Über Anthropotechnik* (Frankfurt 2009) p699ff

INDEX